Symmetry in Special and General Relativity

Symmetry in Special and General Relativity

Special Issue Editor

Jay Tasson

MDPI • Basel • Beijing • Wuhan • Barcelona • Belgrade

Special Issue Editor
Jay Tasson
Carleton College
USA

Editorial Office
MDPI
St. Alban-Anlage 66
4052 Basel, Switzerland

This is a reprint of articles from the Special Issue published online in the open access journal *Symmetry* (ISSN 2073-8994) in 2019 (available at: https://www.mdpi.com/journal/symmetry/special_issues/Symmetry_Special_General_Relativity).

For citation purposes, cite each article independently as indicated on the article page online and as indicated below:

LastName, A.A.; LastName, B.B.; LastName, C.C. Article Title. *Journal Name* **Year**, *Article Number*, Page Range.

ISBN 978-3-03928-094-0 (Pbk)
ISBN 978-3-03928-095-7 (PDF)

Contents

About the Special Issue Editor vii

Preface to "Symmetry in Special and General Relativity" ix

T. H. Bertschinger, Natasha A. Flowers, Serena Moseley, Charlotte R. Pfeifer, Jay D. Tasson and Shun Yang
Spacetime Symmetries and Classical Mechanics
Reprinted from: *Symmetry* **2019**, *11*, 22, doi:10.3390/sym11010022 1

Arnaldo J. Vargas
Overview of the Phenomenology of Lorentz and CPT Violation in Atomic Systems
Reprinted from: *Symmetry* **2019**, *11*, 1433, doi:10.3390/sym11121433 12

Yunhua Ding
Lorentz and CPT Tests Using Penning Traps
Reprinted from: *Symmetry* **2019**, *11*, 1220, doi:10.3390/sym11101220 28

João Alfíeres Andrade de Simões dos Reis and Marco Schreck
Formal Developments for Lorentz-Violating Dirac Fermions and Neutrinos
Reprinted from: *Symmetry* **2019**, *11*, 1197, doi:10.3390/sym11101197 38

Marco Giammarchi
Antimatter Quantum Interferometry
Reprinted from: *Symmetry* **2019**, *11*, 1247, doi:10.3390/sym11101247 54

Lijing Shao
Lorentz-Violating Matter-Gravity Couplings inSmall-Eccentricity Binary Pulsars
Reprinted from: *Symmetry* **2019**, *11*, 1098, doi:10.3390/sym11091098 62

Rui Xu
Modifications to Plane Gravitational Waves from Minimal Lorentz Violation
Reprinted from: *Symmetry* **2019**, *11*, 1318, doi:10.3390/sym11101318 73

Joel Franklin
Symmetric Criticality and Magnetic Monopoles in General Relativity
Reprinted from: *Symmetry* **2019**, *11*, 845, doi:10.3390/sym11070845 82

About the Special Issue Editor

Jay Tasson Assistant Professor of Physics at Carleton College, studies Lorentz symmetry and related topics in a variety of systems. He is a member of the LIGO Scientific Collaboration where he is involved in searches for Lorentz violation as well as detector characterization studies. Jay earned his Ph.D. at Indiana University in 2010 advised by Alan Kostelecky. In the years since, he has taught at several liberal arts colleges culminating in his current appointment at Carleton.

Preface to "Symmetry in Special and General Relativity"

Symmetry has played a crucial role in the development and exploration of special and general relativity. The articles in this Special Issue focus on Lorentz symmetry while also highlighting some of the diverse roles that symmetry plays in these theories.

The first article provides some pedagogical introduction to ideas of symmetry and symmetry violation. The next set of articles focus on Lorentz and CPT symmetry in flat spacetime. They provide, respectively, a review of the phenomenology of Lorentz violation searches in atomic systems, new limits on Lorentz violation from Penning trap experiments, some formal theoretical tools for the study of Lorentz-violating fermions, and presentation of the first evidence of single-antiparticle interferometry. The final three articles shift attention to general relativity, providing, separately, new limits on Lorentz violation from binary-pulsar observations, a study of plane gravitational waves in the presence of Lorentz violation, and an illustration that care must be exercised in using symmetry to extract solutions from the theory.

Jay Tasson
Special Issue Editor

symmetry

MDPI

Article

Spacetime Symmetries and Classical Mechanics

T. H. Bertschinger, Natasha A. Flowers, Serena Moseley, Charlotte R. Pfeifer, Jay D. Tasson * and Shun Yang

Carleton College, One North College St., Northfield, MN 55057, USA; tahbertschinger@gmail.com (T.H.B.); natasha.flowers@sbcglobal.net (N.A.F.); moseleys@carleton.edu (S.M.); pfeiferc@sas.upenn.edu (C.R.P.); shuny@stanford.edu (S.Y.)

* Correspondence: jtasson@carleton.edu

Received: 08 December 2018; Accepted: 24 December 2018; Published: 28 December 2018

Abstract: Physics students are rarely exposed to the style of thinking that goes into theoretical developments in physics until late in their education. In this work, we present an alternative to the traditional statement of Newton's second law that makes theory questions accessible to students early in their undergraduate studies. Rather than a contrived example, the model considered here arises from a popular framework for testing Lorentz symmetry used extensively in contemporary experiments. Hence, this work also provides an accessible introduction to some key ideas in ongoing tests of fundamental symmetries in physics.

Keywords: Lorentz symmetry; rotation invariance; Standard-Model Extension; Noether's theorem

1. Introduction

Is Newton's second law obvious? Some introductory physics students respond in the affirmative. The idea that a force applied to a body results in an acceleration in proportion to the (constant) mass of the body seems to them a clear description of the way nature *must* work. Here, we argue that the answer ought to be "no". We do so by developing a model that contains violations of rotation invariance. Though we develop the model from basic Newtonian-physics considerations, we arrive at the Newtonian limit of a quantum-field-theory based test framework known as the Standard-Model Extension (SME) [1–3]. The SME has been used extensively in searching for violations of Lorentz symmetry (invariance under boosts as well as rotations) in nature [4] with the goal of finding evidence of new physics, such as string theory [5,6].

The construction of physical theories can be thought of as a logical structure, which begins with primitive notations or undefined terms, defines additional concepts from them, and then makes assumptions about how the concepts (defined and undefined) behave. These assumptions are then tested against experimental and observational data to see if the theory so constructed is a description of a physical effect. It is sometimes hard for students and physicists alike to see theories like Newton's laws, which have been around a long time, as fitting this form. This difficulty can make studying the subject feel separate from doing modern science. Newton's laws have also been identified as a particularly challenging example of physical theory [7]. Presenting students with viable alternatives to standard Newtonian theory can help bring the thought processes involved in doing theoretical physics into the undergraduate classroom.

Old ideas in physics can also be difficult to test because physicists have trouble imagining how to do physics without them. Those new to the field of Lorentz-symmetry testing must work to imagine nature without perfect Lorentz symmetry. Rotation invariance is more visual than boost invariance, and it can be readily explored with Newtonian physics. Hence, one can build intuition for symmetry violation with the Newtonian limit of contemporary models of Lorentz-symmetry violation. Testing Lorentz symmetry is an active area of contemporary physics research, and this work provides an accessible introduction to some of its foundational ideas for undergraduates and those new to the field.

In this work, we develop an alternative version of Newton's second law by lifting the assumption of isotropy. In Section 2, we develop the rotation-invariance-violating model from Newtonian considerations, and we address the use of such models in stimulating classroom discussion about the theoretical-physics aspects of Newton's laws. Section 3 introduces the idea of the SME and discusses how our alternative version of Newton's second law fits into it. In Section 4, we explore an example that provides some intuition for how to do physics with the alternative law as well as for how tests of spacetime symmetries are developed. Finally, Section 5 demonstrates the connection between spacetime symmetries and conserved quantities using our alternative Newton's second law as an explicit example.

2. Alternative Newton's Second Laws

A common statement of Newton's second law found in introductory physics courses proceeds as follows: the net force $\vec{F}$ applied to a body is proportional to the acceleration $\vec{a}$ of that body. The proportionality factor, typically taken as constant at this stage, is known as the mass m. The easiest way to imagine an alternative to Newton's second law is to provide a more general form that reduces to the original in some limit. In this section, we consider such examples.

We frame these alternatives in the language above with unaltered force laws such that the simplest limits of our examples may be accessible to students at this level. There are a variety of interpretations of Newton's second law [8]. Hence, some readers might prefer to use $\vec{F} = \frac{d\vec{p}}{dt}$ as the definition of Newton's second law, while recasting the examples to follow as proposed alternative forms for the conserved momentum. Others might wish to interpret the effects we consider as changes to the force laws. We address some of these possibilities in the sections to follow.

Consider first a rotation-invariance-violating (RIV) model with a constant mass. Suppose one applies a given force to a body at rest. One could imagine, for example, that our standard force is defined by stretching a given spring a particular distance. Suppose that the body experiences an instantaneous acceleration a in response to our applied force. Now, suppose that the system is rotated 90 degrees, such that our standard force is applied in a new direction, and in the new configuration a different acceleration, a', results. If such an observation were made, one could imagine modeling it with two Newton's second laws, one for the east–west direction

$$F = ma, \tag{1}$$

and one for the north–south direction

$$F = m'a', \tag{2}$$

with bodies now having two properties, east–west mass m and north–south mass m'. This is a clear violation of rotation invariance. One is then faced with the question of what happens when the system is rotated, not by 90 degrees, but by some other angle. The natural extension is to write Newton's second law in the form

$$F_j = m_{jk}a_k, \tag{3}$$

where Einstein summation convention has been used. In this model, we take m_{jk} as symmetric. While one can consider antisymmetric contributions to m_{jk} here at the level of Newton's second law, such contributions prevent the definition of a kinetic term and hence such models appear to lie outside of action-based theory. We also assume this matrix is invertible. Under these conditions, one finds that forces exerted along three special directions produce accelerations aligned with the force while forces exerted in other directions produce no such alignment. Note that coordinates can always be found that diagonalize the matrix. In these special coordinates, forces aligned with the coordinate axes will produce accelerations aligned with the force and the full model reduces to the original idea introduced in Equations (1) and (2).

Introducing this model to students in mechanics courses produces stimulating discussion that simulates the thinking that happens in theoretical physics. Such discussions can be provoked by asking questions such as, "is this alternative experimentally viable, or has it been ruled out?", "is it internally consistent?", or "how could it be distinguished experimentally from the 'usual' form?". Depending on the level of the course, the presentation can be simplified by using matrix form, and/or using diagonalizing coordinates up front.

Though it has not been confirmed by experiment to date, the RIV model is not pure fiction as it has a clear connection to ongoing efforts in contemporary physics as we discuss in the next section. Such connections can be used to bring recent literature into the classroom. Using notation suggestive of the development to follow, the RIV can be rewritten in the form:

$$F_j = m(\delta_{jk} + 2c_{jk})a_k. \tag{4}$$

Here, an overall factor m equal to 1/3 of the trace of m_{jk} has been pulled out of m_{jk}, and the remaining matrix has been written as the identity (Kronecker delta) plus a traceless matrix $2c_{jk}$. Though we could always choose to write m_{jk} in the form above, this form is particularly convenient when thinking of c_{jk} as a small anisotropic correction to the usual isotropic mass as is typically demanded by existing experimental constraints such as spectroscopy measurements. This form also makes it clear that the model will always be viable for sufficiently small c_{jk}.

The discussion around physical theories and Newton's second law in mechanics courses can be further enhanced by introducing additional examples, which, rather than rotation invariance violation, introduce other modifications. Consider a proportionality factor between the force and acceleration that is a function of some quantity, say the velocity. Hence experimentally, when the same force is applied to a given body (in the lab frame), different accelerations result depending on the velocity the body has at the instant when the force is applied. Consider the following example:

$$F_i = m\left(\gamma\delta_{ij} + \frac{1}{c^2}\gamma^3 v_i v_j\right) a_j, \tag{5}$$

where

$$\gamma = \frac{1}{\sqrt{1 - \frac{v^2}{c^2}}}, \tag{6}$$

and c is a constant with units of velocity. Note that in the limit $v << c$, this alternative would be experimentally indistinguishable from the ordinary case. Hence, for a sufficiently large value of c, this model would remain experimentally viable even if no such velocity dependence were present in nature. Some readers may recognize Equation (5) as a special-relativistic version of Newton's second law [9] common in undergraduate treatments [10]. Introducing this result, or perhaps more appropriately one of its simpler limiting forms such as the case where $\vec{v}$ and $\vec{a}$ are aligned,

$$F = m\gamma^3 a, \tag{7}$$

to students not familiar with special relativity (without saying initially that it's special relativity) produces another model to which the above discussion questions can be applied. Moreover, it demonstrates convincingly that the original notion of Newton's second law is not obvious. For the implication involved in calling something obvious, is that it is obviously right. Since Equation (5) is more correct than the original version, it seems that the original version cannot be obvious. Note that Equation (5) fits the basic form of Equation (3), but the m_{jk} are no longer constant. Note also that, although this match can be made, the directionality in the effective mass in the case of Equation (5) originates from the velocity of the particle in the lab frame, rather than a fundamental violation of rotation invariance. This distinction can be further clarified using the methods to follow in Section 4.

3. The Standard-Model Extension

Among the most fundamental goals of contemporary theoretical physics is the unification of the gravitational interaction with the other three interactions in nature into a single quantum-consistent theory. Several decades ago, the realization that some such unification efforts could generate violations of Lorentz symmetry [5,6] triggered an intense renewed interest in tests of this fundamental spacetime symmetry [4] and the development of a comprehensive test framework for organizing the search [1,2]. This framework is the SME. The idea behind the SME is to add all Lorentz-violating terms to the equations of known physics to form a structure similar to a series expansion about our current best theories. The additional terms can then be sought in experiment. Though the SME expansion is quantum field theory based, the idea is analogous to the addition of c_{jk} to Newton's second law in Equation (4). One could imagine a researcher in Newtonian times proposing Equation (4) as a test framework for deviations for Newtonian physics and seeking c_{jk} in experiments. Such an effort could in principle have discovered either of the models of Section 2. This connection is more than an analogy as the RIV model arises as a subset of the Newtonian limit of the SME. In the remainder of this section, we provide some comments on the connections between the RIV model of Section 2 and the SME, and provide some SME-inspired insights on the RIV model. Some more advanced discussion of the SME is provided in the Appendix A.

In developing the RIV model, we simply imagined the motion of a particle governed by different masses when moving in different directions as modeled by the matrix m_{jk} or equivalently c_{jk}. However, one can visualize objects such as c_{jk} as providing an anisotropy [11] to empty spacetime itself, and the existence of such a condensate of tensors in empty spacetime can be triggered by spontaneous symmetry breaking [12] in analogy with the scalar Higgs field in the Standard Model. The background ovals in Figure 1 illustrate this background condensate.

Figure 1. Block on an inclined plane with gravitational field $\vec{g}$ and background field c_{jk}: (**left**) the systemwith the original coordinates (unprimed) and the observer-rotated coordinates (primed); (**right**) the particle rotated system with the original coordinates. In each case, $\vec{g}$ points toward the center of the Earth.

Via the SME connection, it is straightforward to read off the current experimental limits [4] on how anisotropic the mass in the RIV model can be. The current limits on c_{jk} would permit mass anisotropies (differences in inertia among experiments performed in different directions) at roughly the parts in 10^{17} level in Newtonian experiments in an Earth-based laboratory with conventional macroscopic matter. The least constrained contribution to this number comes from the electron contribution in matter as limited by ion trapping and atomic spectroscopy experiments [13,14]. Proton and neutron contributions are more tightly constrained by a number of tests. Magnetometer experiments [15,16], a type of clock comparison [17–20], are currently the most sensitive.

4. An Example

In this section, we consider one of the most famous examples in physics, the problem of a block on an inclined plane without friction, as a simple example that illustrates some of the interesting features that arise when Newton's second law is generalized to allow rotation-invariance violation [21]. In addition to providing some intuition for the RIV model, the example is of interest for several reasons. It provides what is perhaps the simplest example of some of the conceptual challenges that arise throughout the field of spacetime symmetry testing, some of which are quite foundational to theoretical-physics work more broadly. It is also an interesting example for mechanics students because it forces one to apply a theory without relying on potentially erroneous intuitions.

4.1. Basics

We begin with the simplest case as a starting point for working with nontrivial m_{jk} and as a point of comparison for the extensions to follow. This simplest case is generated when the problem is aligned with the background such that the coordinates that diagonalize m_{jk} are aligned with the plane, as shown in the left-hand diagram of Figure 1, with the x-axis pointing down the plane while the y-axis is perpendicular to the surface. In such coordinates, the gravitational field vector can be expressed as follows:

$$\vec{g} = g(\sin\theta\hat{x} - \cos\theta\hat{y}), \tag{8}$$

where θ is the angle between the surface of the ramp and the horizontal and $g = 9.81\ \mathrm{m/s^2}$ as usual. The gravitational force is then $\vec{F} = m\vec{g}$. Solving for the magnitude of the particle's acceleration down the ramp (x-component here) under the constraint that the acceleration perpendicular to the ramp is zero ($a_y = 0$ here) yields

$$a_x = a_R = (1 - 2c_{xx})g\sin\theta + O(c^2), \tag{9}$$

at leading order in the c_{jk} coefficients, which are known to be small. Note that the only difference from the conventional case is the presence of c_{xx}, and qualitatively the motion remains the same. The particle moves down the ramp in a straight line with constant acceleration.

4.2. Rotations

When a spacetime symmetry is present, transforming the coordinates and the observer's perspective along with them is equivalent to applying the inverse transformation to the items that make up a physical system. When the symmetry is broken, these transformations become inequivalent. Consistent with much of the literature [3], we call the former an observer transformation and the latter a particle transformation. Physical observables should not be affected by observer transformations, while physical violations of spacetime symmetries should be apparent by comparing the results of experiments before and after particle transformations on the experiment. In this section, we apply both transformations in turn and demonstrate that they produce inequivalent results in symmetry-violating models.

First, perform an observer rotation on the original experiment as shown in Figure 1 (left); that is, consider the same problem in new coordinates. Here, we'll consider a rotation by θ such that the gravitational field vector now takes the form

$$\vec{g} = -g\hat{y}'. \tag{10}$$

Schematically, the effective inertial mass will take the form

$$m_{j'k'} = m \begin{pmatrix} 1+2c_{x'x'} & 2c_{x'y'} & 0 \\ 2c_{x'y'} & 1+2c_{y'y'} & 0 \\ 0 & 0 & 1+2c_{z'z'} \end{pmatrix}. \tag{11}$$

Again solving for the acceleration down the ramp subject to the constraint that the perpendicular acceleration is zero yields the components of the acceleration in the new coordinates, which take the form

$$a_{x'} = (1 - 2c_{x'x'}\cos^2\theta - 2c_{y'y'}\sin^2\theta + 4c_{x'y'}\sin\theta\cos\theta)g\sin\theta\cos\theta + O(c^2), \quad (12)$$

$$a_{y'} = -(1 - 2c_{x'x'}\cos^2\theta - 2c_{y'y'}\sin^2\theta + 4c_{x'y'}\sin\theta\cos\theta)g\sin^2\theta + O(c^2). \quad (13)$$

However, this is precisely the acceleration down the ramp found in Equation (9). The match can be made explicit by expressing the components $c_{j'k'}$ in terms of the components c_{jk} via the application of an appropriate rotation matrix:

$$m_{j'k'} = R_{j'j}m_{jk}R_{k'k} \quad (14)$$

and noting that the acceleration is still purely down the ramp with magnitude $a_R = \sqrt{a_{x'}^2 + a_{y'}^2}$. Hence, this example explicitly maintains observer rotation invariance, with both observers agreeing on the outcome of the experiment.

A particle rotation here means that we should pick up and rotate the elements of the experiment (the block, the plane, and the Earth) leaving the coordinates unchanged as shown in Figure 1 (right). Hence, the components of the gravitational field vector change while the components of the background remain the same as we continue to use the original unprimed coordinates that made the mass matrix diagonal. Solving for the motion of the particle in this new rotated configuration yields

$$a_x = (1 - 2c_{xx}\cos^2\theta - 2c_{yy}\sin^2\theta)g\sin\theta\cos\theta + O(c^2),$$
$$a_y = -(1 - 2c_{xx}\cos^2\theta - 2c_{yy}\sin^2\theta)g\sin^2\theta + O(c^2). \quad (15)$$

Here, the magnitude of the acceleration along the ramp is

$$a_R = (1 - 2c_{xx}\cos^2\theta - 2c_{yy}\sin^2\theta)g\sin\theta + O(c^2). \quad (16)$$

Note that this is different from the earlier cases, revealing observable spacetime-symmetry violation. The idea of rotating an experiment illustrated above is a common way of searching for Lorentz violation, most often (though not exclusively) taking advantage of Earth's rotation.

4.3. Discussion

Note that in the above example the form of c_{jk} (as with all vectors and tensors) is coordinate dependent, while physical results are not. This is a general feature. The form of the coefficients for Lorentz violation change under coordinate changes (observer rotations and boosts). Hence, when reporting experimental results, it is convenient to pick a standard frame such that all researchers give the measured coefficients the same name. This standard frame is discussed in Ref. [4]. Occasionally, coordinates can be found that make the symmetry-violating background look special. Such coordinates are sometimes called a preferred frame. The coordinates that diagonalize c_{jk} above can be understood as an example. Often the idea of a preferred frame refers to coordinates in which the physics is rotationally invariant and only boost-symmetry is violated. In general, such preferred frames cannot be found and models in which they exist are special limits of general Lorentz-violating theories.

Some readers might wonder why the mass in the gravitational force law is taken as normal here as opposed to replacing it with m_{jk} as well. There are several related reasons for this choice. First, the mass in Newton's second law and the mass in the law of gravitation reflect two rather different properties in the context of Newtonian physics: the inertia of the body (inertial mass) and the amount of interaction with the gravitational field (gravitational mass). The notion that that inertial mass and gravitational mass are the same is key to the Weak Equivalence Principle [22] and a part of the foundation of General Relativity. Particle-species dependent Lorentz violation typically introduces effective Weak Equivalence Principle violation [23]. Possible violations of the Weak Equivalence

Principle are the subject of much ongoing experimental work [24]. Hence, at this level, independent decisions can be made about their structure.

Second, if the gravitational force contains the same anisotropic effects as the inertial mass, the effects cancel. Hence, even though the equations look more complicated, no observable anisotropy is present in the theory and the laws so written are equivalent to the standard laws. This highlights one of the pitfalls of spacetime-symmetry testing and perhaps of theoretical work more generally: just because a theory is written in a different form this does not necessarily imply a difference in its physical predictions.

Finally, as a limit of the SME, c_{jk} appears only as an effective modification to the inertial mass and not as a modification to the gravitational mass at this level [23]. Another symmetry-violating background in the gravitational sector of the SME quantifies possible Lorentz violation in the gravitational field [25]. In the Newtonian limit, this background generates effective anisotropy of the gravitational mass. It has been shown that a coordinate change can remove the relevant part of c_{jk} from the description of matter, while simultaneously causing it to appear in the gravity sector as an addition to the gravitational field anisotropy [23]. If a special proportionality exists between c_{jk} and the gravitational sector anisotropy, the coordinate redefinition can remove all anisotropic effects from the theory, a result compatible with the cancellation from the Newtonian limit noted above. We note in passing that experimental investigations of gravitational-sector Lorentz violation are also of interest. Recent work by the LIGO, Virgo, Fermi GBM, and INTEGRAL collaborations has placed impressive new constraints via measurements of the speed of gravitational waves [26]. Readers might also have wondered if the spring force law could have been modified instead of the mass during the motivating comments of Section 2. In certain cases, the answer is yes, as these choices are related by a coordinate change analogous to the gravitational force discussion above.

5. Noether's Theorem

Continuous symmetries and conservation laws are intimately connected by Noether's theorem [27]. Since the RIV model violates rotation invariance but maintains spacetime translation invariance, it lacks angular momentum conservation while retaining energy and momentum conservation. In this section, we provide a specific and familiar example that highlights these implications of Noether's theorem.

The system under consideration is a dumbbell composed of a rigid massless rod of length $2l$ and two identical point masses m_{jk} in the RIV model. The system is constrained to the x–y plane with the origin at the midpoint of the system. This set up is a simplified model of the standard "ice-skater-spin" lecture demonstration in which a student spins on a stool holding masses in outstretched arms [28]. For convenience, we work with the Lagrangian formulation. In the RIV model, the kinetic energy T of each mass takes the form

$$T = \tfrac{1}{2} m_{jk} v_j v_k. \tag{17}$$

For the system in question, the Lagrangian, the Hamiltonian, the kinetic energy, and the total energy are all equal. Hence, the Lagrangian for the two-dimensional system can be written

$$L = m\dot{\theta}^2 l^2 (1 + 2c_{xx}\sin^2\theta - 4c_{xy}\sin\theta\cos\theta + 2c_{yy}\cos^2\theta), \tag{18}$$

after implementing the constraints and introducing the plane-polar angle θ in the x–y plane as a generalized coordinate.

Angular momentum is the generalized momentum conjugate to θ, $p_\theta = \frac{\partial L}{\partial \dot{\theta}}$. As usual, the Euler–Lagrange equations

$$\frac{dp_\theta}{dt} = \frac{\partial L}{\partial \theta} \tag{19}$$

imply that p_θ is conserved only if the Lagrangian is independent of θ. Calculation demonstrates that indeed angular momentum is not constant

$$\frac{dp_\theta}{dt} = 2ml^2\dot{\theta}^2\left((c_{xx} - c_{yy})\sin 2\theta - 2c_{xy}\cos 2\theta\right), \tag{20}$$

assuming a nonzero angular speed and at least one nonzero component of c_{jk} in the plane of rotation. We also see on general grounds that energy is conserved since there is no explicit time dependence in L.

To make these conclusions more concrete, we plot angular speed, angular momentum, and energy as a function of time in Figure 2 for a specific choice of parameters. The equation of motion is complicated, but lends itself well to numerical solution. For definiteness and simplicity in this example, we consider the case of $c_{xx} = 0.4$ with all other components of c_{jk} being zero. This large but still perturbative value for c_{jk} provides easily visible results in the plot. Calling the initial angular speed ω_0, we plot the dimensionless angular speed $\Omega = \frac{\dot{\theta}}{\omega_0}$, the dimensionless energy $E = \frac{L}{ml^2\omega_0^2}$, and the dimensionless angular momentum $P_\theta = \frac{p_\theta}{2ml^2\omega_0}$ vs. the dimensionless time $T = \omega_0 t$ for the initial conditions $\theta(0) = 0$, $\Omega(0) = 1$. In conventional physics, the skaters pull their arms closer to the axis of rotation to increase their angular speed and extend their arms to slow their angular speed. Here, we see the perhaps entertaining result that when rotation invariance is violated in this way, the angular speed of the skater varies periodically without changes in the skater's body configuration. We also see explicitly that energy is conserved while angular momentum is not. An animation of these results can be found at https://people.carleton.edu/~jtasson/animations.html.

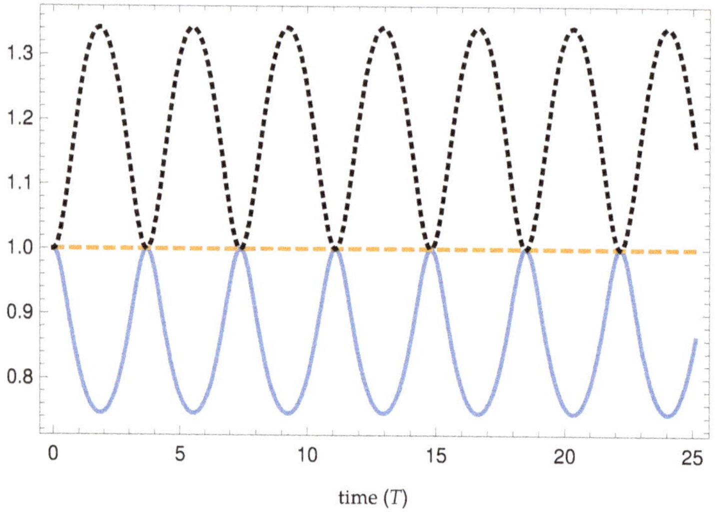

Figure 2. Rigid rotor results vs. dimensionless time: dimensionless angular speed (solid), dimensionless energy (dashed), dimensionless angular momentum (dotted).

6. Conclusions

In this work, we introduced an extension to Newton's second law that permits a generic anisotropic mass. We discussed the use of this model in demonstrating foundational ideas in theoretical physics, and we develop its connection with contemporary efforts to test Lorentz symmetry in the context of the general test framework provided by the SME. We solved the block on the inclined plane as an example to make the ideas concrete and the spinning dumbbell as an illustration of the connection between symmetries and conservation laws. The material presented is useful in teaching theoretical physics ideas in classical mechanics and as an introduction to contemporary efforts to test spacetime symmetries.

Author Contributions: Conceptualization, J.D.T.; formal analysis, T.H.B., N.A.F., S.M., C.R.P., J.D.T., and S.Y.; writing—original draft preparation, J.D.T.; writing—review and editing, T.H.B., N.A.F., S.M., C.R.P., J.D.T., and S.Y.

Funding: This research received no external funding.

Acknowledgments: This work was supported in part by the Carleton College Clinton Ford Physics Research Fund and the Carleton College Towsley Fund. We acknowledge useful conversations on related topics with the following Carleton students: M. Becker, T. Callister, S. Chen, A. Chael, J. Cohen, J. Gais, D. Luo, G. Mo, S. Mehta, F. Mork, R. Skinner, J. Tears, H. Tornabene and S. Wildavsky.

Conflicts of Interest: The authors declare no conflict of interest.

Appendix A

As discussed in Section 3, the RIV model arises as a special limit of the SME framework for testing Lorentz symmetry. In this appendix, we display the SME Lagrange density explicitly such that more advanced readers may see the origin of the object c_{jk} in Equation (4) at this level.

The SME Lagrange density for a free fermion [1,2] can be written as follows:

$$\mathcal{L} = \tfrac{1}{2} i\overline{\psi}(\gamma_\nu - c_{\mu\nu}\gamma^\mu - d_{\mu\nu}\gamma_5\gamma^\mu + \ldots)\overset{\leftrightarrow}{\partial^\nu}\psi - \overline{\psi}(m + a_\mu\gamma^\mu + b_\mu\gamma_5\gamma^\mu + \ldots)\psi. \tag{A1}$$

Here, Greek indices run over the four spacetime dimensions, m is the fermion mass, ψ is the fermion field, $\overline{\psi}$ is the Dirac adjoint fermion field, and the γ^μ are the Dirac matrices. The action of the derivative along with a generic operator Γ^ν is defined by $\overline{\psi}\overset{\leftrightarrow}{\partial^\nu}\Gamma^\mu\psi = \overline{\psi}\Gamma^\mu(\partial_\mu\psi) - (\partial_\mu\overline{\psi})\Gamma^\mu\psi$. The objects a_μ, b_μ, $c_{\mu\nu}$, and $d_{\mu\nu}$ are coefficients for Lorentz violation, which can be identified with the background condensates. They are typically taken as spacetime constants. These coefficients quantify the amount of Lorentz violation in nature and are constrained in experimental and observational searches. The ellipsis contains additional coefficients for Lorentz violation. The object c_{jk}, which has been the primary focus of this work, is the spacial components of the coefficient for Lorentz violation $c_{\mu\nu}$. In the limit that the coefficients for Lorentz violation go to zero, the conventional fermion Lagrange density is recovered. Variation of the Lagrange density in this limit yields the Dirac equation, originally developed as an extension of the Schrodinger equation to provide a description of fermions that incorporated the principles of Special Relativity and quantum mechanics.

Since one could imagine the possibility that Lorentz-violating behavior might be observed in association with just one type of particle, and the goal is a general test framework, the coefficients for Lorentz violation are particle-species dependent. The terms explicitly displayed in Equation (A1) are a part of what is known as the minimal SME: terms involving operators with the same mass dimension [29] as found in the Standard Model and General Relativity. Terms with additional derivatives have also been considered in the fermion [30] sector as well as in the photon [31] and gravity [32] sectors. These nonminimal operators can generate additional effects such as vacuum dispersion. We note in passing that in addition to searches for fundamental symmetry violation, the SME has also been used to constrain and explore fields that can mimic Lorentz violation [33–35].

Several approaches to finding the Newtonian limit of this field theory can be found in the literature [23,36]. In this work we focus on the Newtonian limit generated by the c_{jk} coefficient at leading order in the coefficients for Lorentz violation, as it provides the match to the RIV model as written in Equation (4). In the limit used here, when a macroscopic body that contains many particle species is considered, the effective $c_{\mu\nu}$ coefficient for the body is found to be the mass-weighted sum of the $c_{\mu\nu}$ coefficients for the species contained in the body [23]. Many of the other coefficients have been sought in observations and experiments [4] and have interesting implications in their own right, though their Newtonian implications are beyond our present scope. Additional classical implications can be explored via the Lagrangian formulation [36] as well as a geometric formulation based on Finsler geometry [37,38].

References

1. Colladay, D.; Kostelecký, V.A. Lorentz-violating extension of the Standard Model. *Phys. Rev. D* **1998**, *58*, 116002. [CrossRef]
2. Kostelecký, V.A. Gravity, Lorentz violation, and the Standard Model. *Phys. Rev. D* **2004**, *69*, 105009. [CrossRef]
3. Tasson, J.D. What do we know about Lorentz invariance? *Rep. Prog. Phys.* **2014**, *77*, 062901. [CrossRef] [PubMed]
4. Kostelecký, V.A.; Russell, N. Data tables for Lorentz and CPT violation. *Rev. Mod. Phys.* **2011**, *83*, 11. [CrossRef]
5. Kostelecký, V.A.; Samuel, S. Spontaneous breaking of Lorentz symmetry in string theory. *Phys. Rev. D* **1989**, *39*, 683. [CrossRef]
6. Kostelecký, V.A.; Potting, R. CPT and strings. *Nucl. Phys. B* **1991**, *359*, 545. [CrossRef]
7. Wilczek, F. Whence the Force of F = ma? I: Culture Shock. *Phys. Today* **2004**, *57*, 11. [CrossRef]
8. Saslow, W.M. Accurate physical laws can permit new standard units: The two laws $\vec{F} = m\vec{a}$ and the proportionality of weight to mass. *Am. J. Phys.* **2014**, *82*, 349. [CrossRef]
9. Goldstein, H.; Safko, J.L.; Poole, C.P. *Classical Mechanics*; Pearson: Essex, UK, 2014.
10. Taylor, J.R. *Classical Mechanics*; University Science Books: Sausalito, CA, USA, 2005.
11. Ciftja, O.; Livingston, V.; Thomas, E. Cyclotron motion of a charged particle with anisotropic mass. *Am. J. Phys.* **2017**, *85*, 359. [CrossRef]
12. Kostelecký, V.A.; Bluhm, R. Spontaneous Lorentz Violation, Nambu-Goldstone Modes, and Gravity. *Phys. Rev. D* **2005**, *71*, 065008.
13. Pruttivarasin, T.; Ramm, M.; Porsev, S.G.; Tupitsyn, I.I.; Safronova, M.; Hohensee, M.A.; Haeffner, H. A Michelson-Morley test of Lorentz symmetry for electrons. *Nature* **2015**, *517*, 592. [CrossRef] [PubMed]
14. Hohensee, M.A.; Leefer, N.; Budker, D.; Harabati, C.; Dzuba, V.A.; Flambaum, V.V. Limits on violations of Lorentz symmetry and the Einstein equivalence principle using radio-frequency spectroscopy of atomic dysprosium. *Phys. Rev. Lett.* **2013**, *111*, 050401. [CrossRef] [PubMed]
15. Smiciklas, M.; Brown, J.M.; Cheuk, L.W.; Romalis, M.V. A new test of local Lorentz invariance using ^{21}Ne-Rb-K comagnetometer. *Phys. Rev. Lett.* **2011**, *107*, 171604. [CrossRef]
16. Flambaum, V.V.; Romalis, M.V. Effects of the Lorentz invariance violation in Coulomb interaction in nuclei and atoms. *Phys. Rev. Lett.* **2017**, *118*, 142501. [CrossRef] [PubMed]
17. Kostelecký, V.A.; Lane, C. Constraints on Lorentz violation from clock-comparison experiments. *Phys. Rev. D* **1999**, *60*, 116010. [CrossRef]
18. Altschul, B. Disentangling forms of Lorentz violation with complementary clock comparison experiments. *Phys. Rev. D* **2009**, *79*, 061702. [CrossRef]
19. Kostelecký, V.A.; Vargas, A. Lorentz and CPT tests with clock-comparison experiments. *Phys. Rev. D* **2018**, *98*, 036003. [CrossRef]
20. Lane, C. Using comparisons of clock frequencies and sidereal variation to probe Lorentz violation. *Symmetry* **2017**, *9*, 245. [CrossRef]
21. Bertschinger, T.H.; Flowers, N.A.; Tasson, J.D. Observer and particle transformations. In *CPT and Lorentz Symmetry*; Kostelecký, V.A., Ed.; World Scientific: Singapore, 2014; Volume VI.
22. Nobili, A.M.; Lucchesi, D.M.; Crosta, M.T.; Shao, M.; Turyshev, S.G.; Peron, R.; Catastini, G.; Anselmi, A.; Zavattini, G. On the universality of free fall, the equivalence principle, and the gravitational redshift. *Am. J. Phys.* **2015**, *83*, 39. [CrossRef]
23. Kostelecký, V.A.; Tasson, J.D. Matter-gravity couplings and Lorentz violation. *Phys. Rev. D* **2011**, *83*, 016013. [CrossRef]
24. Touboul, P.; Métris, G.; Rodrigues, M.; André, Y.; Baghi, Q.; Bergé, J.; Boulanger, D.; Bremer, S.; Carle, P.; Chhun, R.; et al. MICROSCOPE Mission: First Results of a Space Test of the Equivalence Principle. *Phys. Rev. Lett.* **2017**, *119*, 231101. [CrossRef] [PubMed]
25. Bailey, Q.G.; Kostelecký, V.A. Signals for Lorentz violation in post-newtonian gravity. *Phys. Rev. D* **2006**, *74*, 045001. [CrossRef]
26. Abbott, B.P.; Abbott, R.; Abbott, T.D.; Acernese, R.; Ackley, K.; Adams, C.; Adams, T.; Addesso, P.; Adhikari, R.X.; Adya, V.B.; et al. Gravitational waves and gamma-rays from a binary Neutron star merger: GW170817 and GRB 170817A. *Astrophys. J. Lett.* **2017**, *848*, L13. [CrossRef]

27. Neuenschwander, D.E. Resource Letter NTUC-1: Noether's Theorem in the undergraduate curriculum. *Am. J. Phys.* **2014**, *82*, 183. [CrossRef]
28. Prigo, R.B.; Reading, M. Quantitative angular momentum experiment on the rotating chair. *Am. J. Phys.* **1977**, *45*, 636. [CrossRef]
29. Tasson, J.D. Antimatter, Lorentz symmetry, and gravity. *JPS Conf. Proc.* **2017**, *18*, 011002.
30. Kostelecký, V.A.; Mewes, M. Fermions with Lorentz-violating operators of arbitrary dimension. *Phys. Rev. D* **2013**, *88*, 096006. [CrossRef]
31. Kostelecký, V.A.; Mewes, M. Electrodynamics with Lorentz-violating operators of arbitrary dimension. *Phys. Rev. D* **2009**, *80*, 015020. [CrossRef]
32. Kostelecký, V.A.; Mewes, M. Testing local Lorentz invariance with gravitational waves. *Phys. Lett. B* **2016**, *757*, 510. [CrossRef]
33. Tasson, J.D. Lorentz violation, gravitomagnetism, and intrinsic spin. *Phys. Rev. D* **2012**, *86*, 124021. [CrossRef]
34. Kostelecký, V.A.; Russell, N.; Tasson, J.D. Constraints on torsion from bounds on Lorentz violation. *Phys. Rev. Lett.* **2008**, *100*, 111102. [CrossRef] [PubMed]
35. Foster, J.; Kostelecký, V.A.; Xu, R. Constraints on nonmetricity from bounds on Lorentz violation. *Phys. Rev. D* **2017**, *95*, 084033. [CrossRef]
36. Kostelecký, V.A.; Russell, N. Classical kinematics for Lorentz violation. *Phys. Lett. B* **2010**, *693*, 443. [CrossRef]
37. Kostelecký, V.A. Riemann-Finsler geometry and Lorentz-violating kinematics. *Phys. Lett. B* **2011**, *701*, 137. [CrossRef]
38. Foster, J.; Lehnert, R. Classical-physics applications for Finsler *b* space. *Phys. Lett. B* **2015**, *746*, 164. [CrossRef]

symmetry

MDPI

Review

Overview of the Phenomenology of Lorentz and CPT Violation in Atomic Systems

Arnaldo J. Vargas

Department of Physics, Loyola University New Orleans, New Orleans, LA 70118, USA; avargas@loyno.edu

Received: 2 November 2019; Accepted: 19 November 2019; Published: 21 November 2019

Abstract: This is an overview of recent publications on the prospects of searching for nonminimal Lorentz-violating effects in atomic spectroscopy experiments. The article discusses the differences in the signals for Lorentz violation in the presence of minimal and nonminimal operators and what systems are more sensitive to certain types of Lorentz-violating operators.

Keywords: Lorentz violation; standard model extension; CPT violation

1. Introduction

Lorentz and CPT symmetry are two of the greatest principles in modern physics. In the last few decades, the exactness of the symmetry has been put into question, and its violation has been pursued as a candidate low energy signal for a quantum theory of gravity. The potential of Lorentz and CPT symmetry as a low energy signal was first proposed following the realization that realistic mechanisms for spontaneous Lorentz- and CPT-breaking in string theory are possible [1,2]. Since then, other studies have suggested that Lorentz- and CPT violation might be low energy signals for several theories beyond the standard model and general relativity such as noncommutative field theory [3–5], loop quantum gravity [6], multiverse scenarios [7], and granular spacetime models [8].

The Standard Model Extension (SME) was introduced as an effective field theory designed to assist in the systematic search for evidence of CPT and Lorentz violation [9,10]. Since the early years of the SME, models for atomic systems in the presence of Lorentz and CPT violation have been proposed [11–15]. Based on these models, experimental bounds on coefficients for Lorentz violation have been reported [16–29]. The original SME, referred to as the minimal SME, only considered the contributions from Lorentz-violating operators of mass dimensions three and four [9,10]. In the last decade, the SME has been extended by considering operators of higher mass dimensions that are called nonminimal operators [30–33]. Relevant to this work is the systematic classification of the Lorentz-violating nonminimal Dirac fermion operators [32]. This classification permitted the study of the prospects of searching for nonminimal Lorentz and CPT violation in atomic spectroscopy experiments. The study of these prospects resulted in three publications [34–36]. The first two publications [34,35] considered light atoms, including exotic atoms such as antihydrogen, positronium, and muonic atoms. The third publication considered heavier atoms that are usually used in high precision spectroscopy experiments or atomic clocks [36]. These publications complement each other, and together, they form a picture of the phenomenology of Lorentz and CPT violation in atomic systems.

This article is intended as a brief overview of the phenomenology of Lorentz and CPT violation in atomic systems based on three recent publications [34–36]. Section 2 is an overview of the perturbative Hamiltonian used in the publications [34–36]. The nonrelativistic coefficients for Lorentz violation are introduced in this section. The next section, Section 3, justifies the use of the perturbation introduced in Section 2. Section 4 discusses the Zeeman-hyperfine transitions of the ground state, which are the most sensitive transitions to Lorentz violation in atomic spectroscopy experiments. Section 5 discusses the prospects for measuring the electron coefficients that do not contribute to the Zeeman-hyperfine transitions of the ground state. Section 6 addresses the problem of testing CPT symmetry in the presence of Lorentz violation. Section 7 discusses differences in the signals for minimal and nonminimal Lorentz-violating terms. Section 8 gives an overview of what systems are more sensitive to certain kinds of Lorentz-violating operators. Finally, we conclude with a brief outlook in Section 9.

2. Classification of the Lorentz- and CPT-Violating Dirac in the Quadratic Lagrange Density for a Dirac Fermion

The first systematic classification of nonminimal Lorentz-violating operators of arbitrary mass dimension was limited to Lorentz-violating photon operators [30]. This work was followed by systematic classifications of nonminimal neutrino operators [31], nonminimal Dirac fermion operators [32], and a more general classification of gauge field theories with nonminimal Lorentz-violating operators [33]. In this article, we will reproduce some of the results presented in [32] as most of the models to be discussed in this review article will be based on the Lorentz violation perturbation terms derived in this reference.

The authors of [32] considered the most general Lorentz-violating Lagrangian density for a free Dirac fermion with flavor w, and it has the form:

$$\mathcal{L} = \tfrac{1}{2}\bar{w}_w(\gamma^\mu i\partial_\mu - m_w + \widehat{\mathcal{Q}}_w)w_w + \text{h.c.}, \tag{1}$$

where w_w is the Dirac fermion field operator, m_w the fermion's mass, and $\widehat{\mathcal{Q}}_w$ is a spinor matrix containing the Lorentz-violating terms. The spinor matrix $\widehat{\mathcal{Q}}_w$ can be represented as the linear combination of the spinor matrices $\gamma^I \in \{I, \gamma^\mu, \gamma_5, \gamma_5\gamma^\mu, \sigma^{\mu\nu}\}$. The linear expansion of $\widehat{\mathcal{Q}}_w$ is assumed to have the form:

$$\widehat{\mathcal{Q}}_w = \sum_I \widehat{\mathcal{Q}}^I_w \gamma_I = \widehat{\mathcal{S}}_w + i\widehat{\mathcal{P}}_w\gamma_5 + \widehat{\mathcal{V}}^\mu_w\gamma_\mu + \widehat{\mathcal{A}}^\mu_w\gamma_5\gamma_\mu + \tfrac{1}{2}\widehat{\mathcal{T}}^{\mu\nu}_w\sigma_{\mu\nu}, \tag{2}$$

where $\widehat{\mathcal{Q}}^I_w \in \{\widehat{\mathcal{S}}_w, \widehat{\mathcal{P}}_w, \widehat{\mathcal{V}}^\mu_w, \widehat{\mathcal{A}}^\mu_w, \widehat{\mathcal{T}}^{\mu\nu}_w\}$ are the expansion coefficients. The hat on top of the coefficients identifies them as functions of the derivative operator $i\partial_\mu$, and they can be expanded as:

$$\widehat{\mathcal{Q}}^I_w = \sum_{d=3}^{\infty} \mathcal{Q}^{(d)I\alpha_1\alpha_2\ldots\alpha_{d-3}}_w i\partial_{\alpha_1} i\partial_{\alpha_2}\ldots i\partial_{\alpha_{d-3}}, \tag{3}$$

where the coefficients $\mathcal{Q}^{(d)I\alpha_1\alpha_2\ldots\alpha_{d-3}}_w$ are the coefficients for Lorentz violation that are assumed to be constant in an inertial reference frame.

The magnitude of a coefficient for Lorentz violation quantifies the degree of the breaking of the Lorentz symmetry. The indexes in the coefficients refer to the properties of the Lorentz violation operators, and Table 1 contains brief explanations of the indices most relevant to the discussion presented in this work. In this discussion also, we will introduce several types of coefficients for Lorentz violation and the terminology used to identify different subsets of the coefficients. For convenience, the terminology needed for this work is collected in Table 2.

The superscript d of the coefficients is the mass dimension of the Lorentz-violating operator that is multiplied by the coefficient $\mathcal{Q}^{(d)I\alpha_1\alpha_2\ldots\alpha_{d-3}}_w$ in Equation (1) after using the expansions in Equations (2) and (3). The expansions in Equations (2) and (3) consider Lorentz-violating operators of arbitrary mass dimension as there is no upper bound on the mass dimension of the operators.

Table 1. Definitions of some of the relevant superscripts and subscripts for the coefficients.

Symbol	Description
d	Mass dimension of the Lorentz-violating operator contracted with the coefficient in the Lagrangian density. Used in effective Cartesian and spherical coefficients.
w	Specifies the flavor of the Lorentz-violating operator contracted with the coefficient in the Lagrangian density. Used in all coefficients.
j	Specifies the rank of the spherical tensor contracted with the coefficient in the one-particle Hamiltonian; $j > 0$. Used in nonrelativistic and spherical coefficients.
m	Specifies the component of the spherical tensor contracted with the coefficient in the one-particle Hamiltonian; $m \in \{-j, -j+1, \ldots, j-1, j\}$. Used in nonrelativistic and spherical coefficients.
n	Specifies the power of the three-momentum when the the one-particle Hamiltonian is expressed in terms of E_0 and $\|\boldsymbol{p}\|$. Used in spherical coefficients; see Equation (8).
k	Specifies the power of the three-momentum when the the one-particle Hamiltonian is expressed in terms of m_w and $\|\boldsymbol{p}\|$. Used in nonrelativistic coefficients; see Equation (10).

Table 2. Terminology used to refer to certain types of coefficients for Lorentz violation.

Terminology	Description	Types of Coefficients
Effective Cartesian	Coefficients for Lorentz-violating operators expressed as Lorentz tensors.	$\mathcal{V}_{w_{\text{eff}}}^{(d)\mu\alpha_1\ldots\alpha_{d-3}}$ $\widetilde{\mathcal{T}}_{w_{\text{eff}}}^{(d)\mu\nu\alpha_1\ldots\alpha_{d-3}}$
Spherical	Coefficients for Lorentz-violating operators expressed as spherical tensors	$\mathcal{V}_{w_{njm}}^{(d)}$ $\mathcal{T}_{w_{njm}}^{(d)(0B)}$, $\mathcal{T}_{w_{njm}}^{(d)(1B)}$
Nonrelativistic	Linear combinations of spherical coefficients of arbitrary mass dimension d.	$\mathcal{V}_{w_{kjm}}^{\text{NR}}$ $\mathcal{T}_{w_{kjm}}^{\text{NR}(0B)}$, $\mathcal{T}_{w_{kjm}}^{\text{NR}(1B)}$
Minimal	Coefficients for minimal operators	Coefficients with $d \leq 4$
Nonminimal	Coefficients for nonminimal operators	Coefficients with $d > 4$
CPT-even	Coefficients for CPT-invariant operators	$\mathcal{V}$-type with even d or c-type $\mathcal{T}$-type with odd d or H-type
CPT-odd	Coefficients for CPT-violating operators	$\mathcal{V}$-type with odd d or a-type $\mathcal{T}$-type with even d or g-type
Spin-dependent	Coefficients proportional to the Pauli matrices in the one-particle Hamiltonian	$\mathcal{T}$-type; or equivalently g-type and H-type
Spin-independent	Coefficients not proportional to the Pauli matrices in the one-particle Hamiltonian	$\mathcal{V}$-type; or equivalently a-type and c-type
Isotropic	Coefficients for rotational scalar Lorentz-violating operators	Spherical or nonrelativistic coefficients with $j = 0$
Anisotropic	Coefficients for Lorentz-violating operators that are not rotational scalars	Spherical or nonrelativistic coefficients with $j > 0$

Starting from the Lagrange density (1), a Lorentz-violating perturbation to the one-particle Dirac Hamiltonian was obtained [32]. The form of the perturbation is:

$$\delta h \;=\; -\frac{1}{E_0}\left[\widehat{\mathcal{V}}^{\nu}_{\text{eff}} + \widetilde{\widehat{\mathcal{T}}}^{0\nu}_{\text{eff}}\frac{\boldsymbol{p}\cdot\boldsymbol{\sigma}}{m_{wi}} + \widetilde{\widehat{\mathcal{T}}}^{i\nu}_{\text{eff}}\left(\sigma_i + p_i\frac{\boldsymbol{p}\cdot\boldsymbol{\sigma}}{(E_0+m_w)m_w}\right)\right]p_\nu, \tag{4}$$

where E_0 is the energy of the fermion, $\boldsymbol{p}$ is the three-momentum of the fermion, and $\boldsymbol{\sigma}$ is the Pauli vector. The terms $\widehat{\mathcal{V}}^{\nu}_{\text{eff}}$ and $\widetilde{\widehat{\mathcal{T}}}^{0\nu}_{\text{eff}}$ can be expressed as polynomials of the components of the four-momentum; see Equations (77) and (79) of [32]. This is similar to the expansion in Equation (3) with the reinterpretation of the operator $i\partial_\mu$ as the one-particle four-momentum operator. The coefficients of the expansion, denoted as $\mathcal{V}_{w_{\text{eff}}}^{(d)\mu\alpha_1\ldots\alpha_{d-3}}$ and $\widetilde{\mathcal{T}}_{w_{\text{eff}}}^{(d)\mu\nu\alpha_1\ldots\alpha_{d-3}}$, are called the effective Cartesian coefficients for Lorentz violation.

The operators contributing to the perturbation (4) can be classified into several categories. The operators multiplied by the $\mathcal{V}$-type coefficients are called spin-independent coefficients as they are independent of the spin degree of freedom. In contrast, operators multiplied by the $\mathcal{T}$-type coefficients are called spin-dependent coefficients. The properties of the operators in (1) under CPT transformation are determined by the mass dimension d of the operator. By convention, different letters are used for the coefficients corresponding to CPT-violating operators and for the ones corresponding to CPT-invariant operators. The spin-independent operators with even mass dimensions are CPT-invariant operators, and the coefficients are c-type coefficients. In contrast, the spin-independent operators with odd mass dimensions are CPT-violating operators, and the coefficients are a-type coefficients. These coefficients are related to the $\mathcal{V}$-type coefficients by:

$$\mathcal{V}_{w_{\text{eff}}}^{(d)\mu\alpha_1\ldots\alpha_{d-3}} = \begin{cases} -a_{w_{\text{eff}}}^{(d)\mu\alpha_1\ldots\alpha_{d-3}} & \text{if } d \text{ is odd} \\ +c_{w_{\text{eff}}}^{(d)\mu\alpha_1\ldots\alpha_{d-3}} & \text{if } d \text{ is even} \end{cases}. \tag{5}$$

The spin-dependent terms can also be divided into CPT-invariant and CPT-violating terms. The relation between the $\mathcal{T}$-type coefficients and the other set of coefficients is given by:

$$\widetilde{\mathcal{T}}_{w_{\text{eff}}}^{(d)\mu\nu\alpha_1\ldots\alpha_{d-3}} = \begin{cases} -\widetilde{H}_{w_{\text{eff}}}^{(d)\mu\nu\alpha_1\ldots\alpha_{d-3}} & \text{if } d \text{ is odd} \\ +\widetilde{g}_{w_{\text{eff}}}^{(d)\mu\nu\alpha_1\ldots\alpha_{d-3}} & \text{if } d \text{ is even} \end{cases}, \tag{6}$$

where H-type coefficients correspond to CPT-invariant operators and the g-type coefficients to CPT-violating operators.

The perturbation Hamiltonian (4) can be expressed in momentum-space spherical coordinates instead of Cartesian coordinates. The three-momentum $\boldsymbol{p}$ is the product of its magnitude $|\boldsymbol{p}|$ and direction $\hat{\boldsymbol{p}}$. The unit vector in the direction of the three-momentum can be represented as a function of the polar and azimuthal angles as $\hat{\boldsymbol{p}} = (\sin\theta\cos\phi, \sin\theta\sin\phi, \cos\theta)$. The direction of the Pauli vector can be indicated in terms of the direction of the three-momentum by introducing a helicity basis with unit vectors $\hat{\boldsymbol{e}}_\pm = (\hat{\boldsymbol{\theta}} \pm i\hat{\boldsymbol{\phi}})/\sqrt{2}$ and $\hat{\boldsymbol{e}}_r = \hat{\boldsymbol{p}}$. After these changes, the Hamiltonian has the generic form:

$$\delta h = h_{w0} + h_{wr}\boldsymbol{\sigma}\cdot\hat{\boldsymbol{e}}_r + h_{w+}\boldsymbol{\sigma}\cdot\hat{\boldsymbol{e}}_- + h_{w-}\boldsymbol{\sigma}\cdot\hat{\boldsymbol{e}}_+. \tag{7}$$

The explicit expressions for the terms h_{w0}, h_{wr}, h_{w+}, and h_{w-} can be found in Equations (85) and (87) in [32]. As an example, consider the expression for h_{w0},

$$h_{w0} = \sum_{d=3}^{\infty}\sum_{n=0}^{d-2}\sum_{j}\sum_{m=-j}^{j} E_0^{d-3-n}|\boldsymbol{p}|^n\, Y_{jm}(\hat{\boldsymbol{p}})\, \mathcal{V}_{w_{njm}}^{(d)}, \tag{8}$$

where the sum over j is restricted to $j \geq 0$ and $j \in \{n, n-2, n-4, \ldots\}$. The coefficients $\mathcal{V}_{w_{njm}}^{(d)}$ are called the spherical coefficients for Lorentz violation. The spherical coefficients for Lorentz violation are linear combinations of the effective Cartesian coefficients for Lorentz violation. The relation between the two sets of coefficients is explained in detail in Section IV of [32]. In the Equation (8), the symbol

$Y_{jm}(\hat{\boldsymbol{p}})$ represents the spherical harmonics and the subscripts j and m of the spherical coefficients label the indices of the corresponding spherical harmonic. The index d is the mass dimension of the operator, and the index k is the power of the magnitude of the three-momentum. The relation between the indices j, d, and n for all the different types of coefficients is summarized in Table III of [32].

The Hamiltonian (7) is only valid at linear order on the coefficients for Lorentz violation. At this order, the energy E_0 can be assumed to be the energy for a free fermion given by $E_0 = \sqrt{|\boldsymbol{p}|^2 + m_w^2}$. In nonrelativistic systems, the ratio $\boldsymbol{p}/m_w$ is a small number that can be used to expand the energy as a Taylor series. Using the binomial formula, we have that:

$$E_0 = m_w \sqrt{1 + \left(\frac{|\boldsymbol{p}|}{m_w}\right)^2} = m_w \sum_{k=1}^{\infty} \binom{\frac{1}{2}}{k} \left(\frac{|\boldsymbol{p}|}{m_w}\right)^{2k}, \tag{9}$$

where $\binom{j}{k}$ is the binomial coefficient. Using this formula, we can express Equation (8) as:

$$h_{w0} = -\sum_{kjm} |\boldsymbol{p}|^n \, {}_0Y_{jm}(\hat{\boldsymbol{p}}) \mathcal{V}_{w\,kjm}^{\mathrm{NR}}, \tag{10}$$

where $\mathcal{V}_{w\,kjm}^{\mathrm{NR}}$ is the linear combination of all the spherical coefficients $\mathcal{V}_{w\,njm}^{(d)}$ that are proportional to the same power of $|\boldsymbol{p}|$ after replacing the energy in Equation (8) in terms of $|\boldsymbol{p}|$ by using Equation (9). The index n in Equation (8) is the power of $|\boldsymbol{p}|$ when the Hamiltonian was represented as a function of the energy and three-momentum, and it is different from the index k in Equation (10) that corresponds to the power of $|\boldsymbol{p}|$ after replacing the energy using Equation (9).

The coefficients $\mathcal{V}_{w\,kjm}^{\mathrm{NR}}$ are called the nonrelativistic coefficients and are the observable coefficients in most nonrelativistic experiments. The term observable effective coefficients means that the Lorentz-violating shift to the observables in nonrelativistic experiments can be expressed as linear combinations of the nonrelativistic coefficients. The nonrelativistic coefficients are defined in Equations (111) and (112) of [32]. For instance, consider the definition of $\mathcal{V}_{w\,kjm}^{\mathrm{NR}}$,

$$\mathcal{V}_{w\,kjm}^{\mathrm{NR}} = \sum_{d} m_w^{d-3-k} \sum_{q \le k/2} \binom{(d-3-k+2q)/2}{q} \mathcal{V}_{w\,(k-2q)jm}^{(d)}. \tag{11}$$

The nonrelativistic coefficients are the linear combination of coefficients for Lorentz violation of arbitrary mass dimension multiplied by powers of the fermion's mass m_w. The mass dimension of the nonrelativistic coefficients can be determined with some basic dimensional analysis. If the operator multiplied by the coefficient $\mathcal{V}_{w\,njm}^{(d)}$ has mass dimension d, then the coefficient has mass dimension $4 - d$. The mass dimension of the nonrelativistic coefficient $\mathcal{V}_{w\,kjm}^{\mathrm{NR}}$ is the mass dimension of $\mathcal{V}_{w\,njm}^{(d)}$ multiplied by the mass dimension of m_w^{d-3-k}. Putting the pieces together, we can conclude that the mass dimension of $\mathcal{V}_{w\,kjm}^{\mathrm{NR}}$ is equal to $1 - k$.

In many nonrelativistic experiments, it is impossible to distinguish between the spherical coefficients that contribute to the same nonrelativistic coefficient [34–36]. For that reason, Lorentz violation effects in atomic systems usually result in bounds on the nonrelativistic coefficients for Lorentz violation. Exceptions to this rule are Lorentz-violating models that consider contributions due to the electromagnetic fields [37] or boost effects [35,36]; see Section 8.

3. Hierarchy and the Lorentz-Violating Perturbation

The Lorentz-violating corrections to the free propagation of the electron and the proton in the hydrogen atom are expected to be responsible for the dominant Lorentz- and CPT-violating effects if we consider all the possible Lorentz-violating operators [12,35,38]. The previous statement needs some clarification. In the context of models for Lorentz violation in atomic systems, there are two kinds of small parameters. The first kind is the expansion parameters used to obtain corrections to the atomic energies using perturbative methods. Examples of these parameters are the ratio $|\boldsymbol{p}|/m_w$ between the magnitude of the three-momentum and the mass of the electron, the ratio m_w/M between the mass of the electron and the mass of the nucleus, and the fine structure constant α. These parameters introduce a hierarchy on the atomic corrections.

The second kind of small parameter is the coefficients for Lorentz violation. The coefficients for Lorentz violation are considered small parameters to be measured, but before measuring them, we cannot compare two coefficients for Lorentz violation. For example, we cannot tell which one of the following dimensionless terms $c_{w200}^{(4)}$, $c_{w210}^{(4)}$ or $m_w\, a_{w200}^{(5)}$ is greater. A common practice is not to assume a hierarchy between the coefficients for Lorentz violation in the absence of experimental bounds. We consider all the coefficients to be independent of each other. We also usually consider only linear contributions due to the coefficients for Lorentz violation; therefore, any hierarchy on the perturbative corrections is due to the atomic expansion parameters. For each coefficient for Lorentz violation, we have a perturbative series that has a similar hierarchy as the usual atomic corrections in the absence of Lorentz violation. For example, the Lorentz-violating contributions to the energy shift proportional to the same coefficient can be classified or ranked in terms of the nonrelativistic expansion that is the expansion on the small parameter $|\boldsymbol{p}|/m_w$.

To illustrate the idea, we need to study the form of the nonrelativistic expansion for a free Dirac fermion. Using Equation (9), we obtain:

$$E_0 = \sqrt{|\boldsymbol{p}|^2 + m_f^2} \simeq m_f\Big(1 + \frac{1}{2}\Big(\frac{|\boldsymbol{p}|}{m_w}\Big)^2 - \frac{1}{8}\Big(\frac{|\boldsymbol{p}|}{m_w}\Big)^4 + \ldots\Big). \tag{12}$$

The contributions at different orders in the expansion have the generic form $m_w(|\boldsymbol{p}|/m_w)^n$. Even in the case of a Dirac fermion in the presence of an external electromagnetic field, we can expand the Hamiltonian in terms of the small parameter $|\boldsymbol{p}|/m_w$ using the Foldy–Wouthuysen transformation [39]. In the particular case of the hydrogen atom, the Coulomb-potential term appears at the first-order in the nonrelativistic expansion, but it is suppressed by a factor of the fine structure constant α that makes the Coulomb term of the same size as a second-order term such as the nonrelativistic kinetic energy $|\boldsymbol{p}|^2/2m_w$.

Consider the term $E_0|\boldsymbol{p}|^2 c_{w200}^{(6)}$ that contributes to the one-particle Hamiltonian (8). If we want to determine the dominant contribution from the coefficient $c_{w200}^{(6)}$ to the energy shift, we can use the nonrelativistic expansion of the energy and get:

$$E_0|\boldsymbol{p}|^2 c_{w200}^{(6)} \simeq m_f\Big(1 + \frac{1}{2}\Big(\frac{|\boldsymbol{p}|}{m_w}\Big)^2 - \frac{1}{8}\Big(\frac{|\boldsymbol{p}|}{m_w}\Big)^4 + \ldots\Big)|\boldsymbol{p}|^2 c_{w200}^{(6)}. \tag{13}$$

Using this result, we can recognize that a term of the form:

$$m_f|\boldsymbol{p}|^2\, c_{w100}^{(6)} = m_w\left(\frac{|\boldsymbol{p}|}{m_w}\right)(\,|\boldsymbol{p}|\, a_{w200}^{(6)}\,), \tag{14}$$

is a zero-order term in the nonrelativistic expansion, and it is the dominant contribution from the coefficient $c_{w200}^{(6)}$. In the context of the nonrelativistic expansion of the Lorentz-violating perturbation, this term is a large term of the order of the rest energy of the particle, and it is greater than any Lorentz-violating term proportional to $c_{w100}^{(6)}$ that is produced by the electromagnetic interaction in the atom. However, its contribution to the atomic energy is really small because it is proportional to a coefficient for Lorentz violation. The crucial point is that we know that any term that is proportional to both the coefficient and an interaction term such as the Coulomb potential will be smaller than this term.

We can also consider the term of the form:

$$\frac{|\boldsymbol{p}|^4}{m_w} c_{w100}^{(6)} = m_w \left(\frac{|\boldsymbol{p}|}{m_w}\right)^2 \left(|\boldsymbol{p}|^2 c_{w200}^{(6)} \right) \tag{15}$$

and recognize that it is a second-order term in the nonrelativistic expansion; it is not the dominant contribution from the coefficient $c_{w200}^{(6)}$, and it can contribute at the same order as a Coulomb potential term that is proportional to the same coefficient; for that reason, in order to study this term, we must consider the Lorentz-violating electromagnetic interaction terms [12,35,38]. Fortunately, in practice, we can ignore this term and only consider the term $m_f |\boldsymbol{p}|^2 c_{w100}^{(6)}$ that dominates over the Lorentz-violating terms that contain electromagnetic interactions.

Going back to the statement at the beginning of this section, the dominant contribution to the atomic spectrum is obtained by considering only the dominant free-propagation corrections to the proton and the electron for each coefficient. The implication is that it is enough to consider the perturbative Hamiltonian (7) in order to study the dominant Lorentz-violating effects in the spectrum of an atom [35,36].

4. Hyperfine Transitions and Anisotropic Terms

The best limits on Lorentz-violating operators obtained from atomic spectroscopy experiments are from hyperfine transitions of the ground state [34–36]. In the standard atomic theory, effects that depend on the total angular momentum of the atom, such as the hyperfine structure, are suppressed. For this reason, in general, hyperfine structure transitions have lower frequencies than gross structure transitions. On the other hand, many of the dominant Lorentz-violating terms are anisotropic, and their expectation values depend on the atomic total-angular-momentum quantum number F. For example, consider the term $m_w g_{w010}^{(4)(0B)} \sigma \cdot \hat{\boldsymbol{p}}\, Y_{10}(\hat{\boldsymbol{p}})$. This is the dominant contribution of the coefficient $g_{w010}^{(4)(0B)}$ to the perturbation Hamiltonian as the other contributions are suppressed by powers of $|\boldsymbol{p}|/m_w$. Because this term depends on the spin expectation value, it does contribute to hyperfine structure transitions [34–36]. What makes this kind of term special is that its contribution has the same size as the gross structure or hyperfine structure transitions. However, because the hyperfine transitions are usually more sensitive to smaller frequency shifts than gross structure transitions, then the hyperfine structure transitions are more sensitive to the coefficient $g_{w010}^{(4)(0B)}$ than other types of transitions [34–36].

At first-order in perturbation theory, the anisotropic terms in the Lorentz-violating Hamiltonian affect the spectrum in a fashion that is analogous to the presence of small external electric and magnetic fields. For example, some of the leading-order Lorentz-violating energy shifts have a structure that resembles the Zeeman and Stark effects [12–15,34–36]. This is a challenge because transitions that are insensitive to Zeeman or Stark effects may also be insensitive to these Lorentz-violating effects. To understand this statement, we need to understand what are the common tests for Lorentz violation in atomic systems.

The most common tests for Lorentz violation are sidereal and annual variations of the transition frequency [16–23]. The idea behind these tests is to compare the transition frequency of the atom at different velocities and orientations relative to a fixed inertial reference frame. For convenience, the Sun-centered frame is used as the fixed reference frame [40]. The best approach to control the orientation of the atom is to introduce an external magnetic field in the z-direction in the instantaneous laboratory frame. Because of the presence of the magnetic field, the stationary states of the system are quantum states of the z-component of the total-angular-momentum F_z relative to the laboratory frame. As the applied magnetic field rotates with the Earth, the stationary states are rotated adiabatically around the Sun-centered frame, and we can test the rotational symmetry of the atomic spectrum. Similarly, the velocity of the atoms changes as the atoms are accelerated in the Sun-centered frame due to the rotation of the Earth around its axis and the motion of the Earth around the Sun. In this scenario, the Lorentz-violating terms appear as small corrections to the Zeeman levels that depend on the annual and sidereal time [12–15,34–36].

If we are forced to use applied magnetic fields, then we want to reduce the uncertainty due to the magnetic fields. A common method is to use transitions insensitive to the linear Zeeman effect. Examples of this kind of transition are the clock transitions in hydrogen masers and cesium atomic fountain clocks. This is a bad idea in the context of Lorentz violation. Whatever makes these transitions insensitive to the linear Zeeman effect also makes them insensitive to linear effects due to other uniform anisotropic external fields such as the anisotropic Lorentz-violating background fields [35,36]. In other words, these transitions are insensitive to the dominant CPT- and Lorentz-violating effects. Still, there is one advantage of having transitions that are insensitive to the dominant Lorentz-violating terms. We measure a frequency by comparing it to another frequency. We need to know if the Lorentz-violating model predicts any variation in the reference frequency in order to search for time variations of a transition frequency [13]. Using the perturbation (7), we know that the hydrogen maser and the cesium standards are insensitive to the dominant Lorentz-violating effects and are good reference frequencies for time variation studies of transition frequencies [35,36].

Other methods used to reduce the uncertainty due to the magnetic field, such as averaging over Zeeman pairs, can also eliminate contributions due to the anisotropic Lorentz-violating terms [36]. For example, consider the measurement of the hyperfine transition of the ground state of antihydrogen [41]. In the experiment, two frequencies were averaged to suppress the contribution due to the magnetic field. This process also eliminated the contributions from the dominant CPT-violating terms. In other words, the sensitivity of the measurement to CPT violation is suppressed compared to other kinds of tests that could be done using the same system. A method that can be used to eliminate the magnetic field without eliminating the contribution of the Lorentz-violating terms is to extrapolate the frequency to the zero-magnetic-field value [36]. The dominant CPT-violating terms are independent of the magnitude of the magnetic field, and they will contribute to the extrapolated zero-field frequency. Another method that has been proposed is to compare the σ and π_1 antihydrogen transitions [42]. The σ transition is insensitive to the dominant CPT-violating terms, and it can be used as a reference frequency for searching for a sidereal variation of the Lorentz violation-sensitive π_1 transition [35].

Averaging over Zeeman pairs does not always cancel all the contributions due to Lorentz violation. For example, a Lorentz symmetry test with cesium fountain clocks cannot use the standard clock transition as it is insensitive to the dominant Lorentz-violating terms. Time variation frequency studies with cesium fountain clocks were done using an averaged pair of hyperfine-Zeeman transitions [22]. The process used to eliminate the linear Zeeman shift also canceled the contributions from the g-type and H-type spin-dependent coefficients, but it allowed contributions from a-type and c-type spin-independent ones [22,36]. The most successful method for eliminating the linear Zeeman effect without eliminating the Lorentz-violating terms has been the use of comagnetometers [18,19,36]. The nonrelativistic g-type and H-type coefficients for Lorentz violation with $j = 1$ produce small corrections to the Zeeman levels; however, the corrections are not proportional to the gyromagnetic ratios and are by the method used to eliminate the linear Zeeman shift in the comagnetometer.

The only spatially isotropic terms that can contribute to the Lorentz-violating shift to the atomic spectrum are spin-independent operators that depend only on the magnitude of the three-momentum [34–36]. These isotropic terms do not contribute to Zeeman-hyperfine transitions, and for that reason, all the terms that contribute to these transitions are anisotropic. The Lorentz-violating frequency shift for the Zeeman-hyperfine transitions depends on the orientation of the magnetic field and the boost velocity of the instantaneous laboratory frame relative to the Sun-centered frame. Only considering the rotation of the instantaneous laboratory frame due to the rotation of the Earth is not enough to impose bounds on all the coefficients for Lorentz violation that contribute to the transition frequencies. Models for space based experiments such as for the Atomic Clock Ensemble in Space (ACES) [43] and for experiments on turntables have been considered to impose bounds on a greater set of coefficients for Lorentz violation [35,36].

5. Isotropic Terms and Optical Transitions

The isotropic term in the laboratory frame has the form $\mathcal{V}_{f\,k00}^{\text{NR,lab}}|\boldsymbol{p}|^k$, where the superscript "lab" is a reminder that these coefficients are not constant and uniform because the laboratory frame is not an inertial reference frame. The isotropic term does not contribute to the frequency shift for the hyperfine-Zeeman transitions of the ground state, and for that reason, it cannot be measured using the experiments mentioned in Section 4. This term does contribute to any gross structure transition such as optical transitions. The best candidates to study the isotropic term are optical transitions such as the 1 s–2 s transition in hydrogen [35] or clock transitions used in optical clocks [36].

The isotropic term in the laboratory frame is independent of the orientation of the magnetic field, and it is not canceled by any process that cancels the contributions from anisotropic external fields [36]. The implication is that the isotropic term can be studied using optical clocks without requiring the optical clock to operate in a different way than usual. The drawback is that the isotropic term is insensitive to changes in the orientation of the laboratory frame that is the dominant signal for Lorentz violation. However, it is sensitive to boost effects, which are suppressed by a factor of 10^{-4} compared to the dominant rotation effects.

An isotropic coefficient for Lorentz violation in the laboratory frame can be expressed in the Sun-centered frame as:

$$\mathcal{V}_{f\,k00}^{\text{NR,lab}} = \mathcal{V}_{f\,k00}^{\text{NR,Sun}} + \beta_{\oplus} f_{ann}(T) + \beta_L f_{sid}(T), \qquad (16)$$

where $\mathcal{V}_{f\,k00}^{\text{NR,Sun}}$ is the isotropic coefficient in the Sun-centered frame, T is the time in the Sun-centered frame, $\beta_{\oplus} = 10^{-4}$ is the orbital speed of the Earth, and $\beta_L = 10^{-6}$ is the rotational speed of the Earth at the Equator. The function $f_{ann}(T)$ is a linear combination of coefficients for Lorentz violation with terms that vary with the first harmonic of the annual frequency and $f_{sid}(T)$ the same, but the terms vary with the first harmonics of the sidereal frequency. The explicit expression for Equation (16) can be found in Equation (63) in [35]. The best way to impose constraints on the coefficients that contribute to $f_{ann}(T)$ and $f_{sid}(T)$ is by searching for annual and sidereal variations of the optical transitions in the first harmonic of the sidereal and annual frequency. The first term in Equation (16) produces a constant shift that will be the same independent of wherever on the surface of the Earth the experiment was done. This constant shift cannot be constrained by studying the time variation of the transition frequency under consideration. However, the isotropic term in the Sun-centered frame has been constrained by comparing the 1 s–2 s transitions frequency of hydrogen and antihydrogen [36] or by comparing the experimental and theoretical values for the 1 s–2 s transitions in positronium [35] and muonium [34].

6. The Problem of Testing CPT Symmetry Using Different Frames

A breaking of CPT symmetry implies Lorentz violation in interacting local field theories [44]. This result is also observed in the non-gravitational sector of the SME, where all the local CPT-violating terms that can be added to the Lagrangian density also break Lorentz symmetry [9,10]. If we expect CPT violation to emerge as small corrections to the standard model of particle physics, then we expect CPT violation to be accompanied by Lorentz violation. This observation implies that CPT tests that compare the properties of a system and its CPT counterpart must be conducted in the same laboratory frame. Otherwise, Lorentz-violating effects that are not CPT-violating effects might be responsible or might cancel any discrepancy between the two systems [37].

Even if the measurement of a system and its CPT counterpart is done in different reference frames, we could use these results to test CPT symmetry by using a model for Lorentz violation. We can use the model to transform the results from one frame to the other keeping track of all the Lorentz-violating effects. In this case, the validity of the CPT test will be limited as it depends on the particular model for CPT and Lorentz violation used. For example, consider the recent comparison between the value of the 1 s–2 s transition in hydrogen [45] and antihydrogen [46]. As these two values were measured in different reference frames, in principle, we should not compare the values without considering how

Lorentz violation could impact these results. Using the Lorentz- and CPT-violating corrections for the 1 s–2 s transition in hydrogen presented in [35], a model for comparing the two measurements of the 1 s–2 s transition was developed [36]. The model considered only the isotropic contribution in the Sun-centered frame to the frequency difference between the 1 s–2 s transition in hydrogen and antihydrogen; see Section 5. In other words, even if the frequencies were measured in distinct reference frames, there are corrections to the frequency difference that are independent of the frames used in the measurements, and these terms correspond to the constant term in Equation (16). To justify this model partially, some of the anisotropic or frame-dependent contributions to the frequency difference can be disregard using results from time variation studies of transition frequencies in hydrogen [16,20]. More time variation studies in hydrogen and antihydrogen are needed to justify experimentally the absence of many of the anisotropic terms that were not considered in the model. Fortunately, some of these time variation studies are expected to happen soon [42]. The approach used to create the model for the frequency difference between the 1 s–2 s transition in hydrogen and antihydrogen cannot be replicated for the Zeeman-hyperfine transitions of the ground state as all the terms that contribute to these transitions are anisotropic in the Sun-centered frame. Measurements of hyperfine transition frequencies of the ground state for hydrogen and antihydrogen in the same location are pursued to avoid any contributions from CPT-invariant Lorentz-violating operators [42].

7. Difference in the Signals for Minimal and Nonminimal Lorentz-Violating Terms

For experiments in laboratories on the surface of the Earth, the minimal Lorentz-violating operators could produce sidereal variations in the first and second harmonic of the sidereal frequency [11–14]. In the context of atomic spectroscopy experiments, we can understand this result from the following observations. The minimal Lorentz violation a-type and c-type coefficients are contained in the nonrelativistic coefficients $\mathcal{V}_{w_{kjm}}^{\text{NR}}$ with $j \leq 2$, and similar relations hold for the spin-dependent terms [32]. We can break the time-varying transition frequency shift in the Sun-centered frame in terms of harmonics of the sidereal frequency [35,36]. If we ignore boost effects, we can break the Sun-centered-frame transition frequency shift $\delta\nu$ in the following way,

$$\delta\nu = \sum_{m=0}^{\infty} \Big(A_m \cos m\omega_\oplus T + B_m \sin m\omega_\oplus T \Big), \tag{17}$$

where $\omega_\oplus \simeq 2\pi/(23\text{ h } 56\text{ m})$ is the sidereal frequency and T is the time in the Sun-centered frame. The amplitudes A_m and B_m of the mth-harmonics are linear combinations of the coefficient $\mathcal{V}_{w_{kjm}}^{\text{NR}}$ and $\mathcal{V}_{w_{kj(-m)}}^{\text{NR}}$ [35,36]. In other words, the absolute value $|m|$ of the index m of the coefficient for Lorentz violation in the Sun-centered frame indicates the harmonic of the sidereal frequency that contributes together with the coefficient $\mathcal{V}_{w_{kjm}}^{\text{NR}}$ in the frequency shift. The absolute value of the index m is related to the index j by $0 \leq |m| \leq j$ and for the minimal operators $0 \leq |m| \leq 2$. As expected, the minimal operators can only produce sidereal variations in the first and second harmonic of the sidereal frequency.

In principle, the nonminimal Lorentz-violating operators can produce variations with all harmonics of the sidereal frequency. However, nonminimal coefficients with an index j cannot contribute to the energy shift of every atomic energy level. The maximum value of the index j that can contribute to the energy shift depends on the angular momentum quantum numbers of the energy level [35,36]. For instance, for the ground state of hydrogen, the angular moment quantum numbers are $L = 0$ for the orbital angular momentum, $J = 1/2$ for the total electron angular momentum, and $F = 0$ or $F = 1$ for the total atomic angular momentum. Based on the angular momentum quantum numbers, we can conclude that only spin-independent terms with $j = 0$ and spin-dependent ones with $j \leq 1$ can contribute to the energy shift. A consequence of this observation is that even in the presence of nonminimal terms, we should expect only first-harmonic sidereal variations of Zeeman-hyperfine transitions of the ground state of hydrogen or the 1 s–2 s transition in hydrogen. These are the same signals predicted by the minimal SME, and for that reason, experimental constraints on these signals already existed and were used to impose bounds on nonminimal coefficients [34–36].

In the minimal case, there was no advantage in considering transitions involving energy levels with high angular momentum quantum numbers. However, the only way to use an atomic spectroscopy experiment to search for Lorentz-violating operators with a high value of j is by using transitions that involve high angular momentum states. In general, a transition could be millions of times more sensitive to Lorentz violation than another transition, but because that more sensitive transition only involves low angular momentum states, it will be sensitive to a small set of Lorentz-violating operators, and if the less sensitive transition involves high angular momentum states, it can provide the best bounds on coefficients for Lorentz violation on nonminimal operators that cannot be studied with the more sensitive transition; see Section 8.

Another difference in the phenomenology of atomic spectroscopy in the presence of minimal and nonminimal Lorentz violation is that the nonminimal terms depend on a higher power of the three-momentum, and this means that the number of transitions that can be affected by Lorentz violation increased significantly compared to the minimal case [35]. Furthermore, this means that the sensitivity of the experiment to the nonminimal coefficients will be dependent on the expectation values of the momentum, and that will make some systems more sensitive to some nonminimal operators than others as is the case with muonic hydrogen and muonium, as discussed in [34].

8. Best Bounds on and Prospects for Coefficients for Lorentz Violation from Spectroscopy Experiments

Table 3 contains the best bounds on the nonrelativistic spin-dependent coefficients for Lorentz violation. The first column in the table specifies the type of nonrelativistic coefficient, and the other columns specify the best bounds on the electron, neutron, proton, and muon coefficients. A time variation study of hyperfine-Zeeman transition frequencies of the ground state of hydrogen is responsible for the best bounds on nonminimal electron coefficients obtained in atomic spectroscopy experiments [16,35]. The bounds obtained on the coefficients $g_{e011}^{\mathrm{NR}(0B)}$, $H_{e011}^{\mathrm{NR}(0B)}$, $g_{e011}^{\mathrm{NR}(1B)}$, and $H_{e011}^{\mathrm{NR}(1B)}$ are in the order of 10^{-27} GeV [35]. The superscript e in the coefficients means that these coefficients correspond to electron operators. Technically, this experiment also has the best bounds on nonrelativistic proton coefficients, but better bounds on proton coefficients might be obtained by just replacing the nuclear model used in [36].

An experiment using a ^{3}He-^{129}Xe comagnetometer imposed limits of the order of 10^{-33} GeV on Lorentz-violating operators in the nucleon sectors of the SME [18,36]. To assign these bounds to proton or neutron Lorentz-violating operators, we need to use a nuclear model. A simple nuclear model assumes that only the neutron operators contribute to the Lorentz-violating frequency shift, and using this simplistic model, bounds of the order of 10^{-33} GeV on the neutron coefficients $g_{n011}^{\mathrm{NR}(0B)}$, $H_{n011}^{\mathrm{NR}(0B)}$, $g_{n011}^{\mathrm{NR}(1B)}$, and $H_{n011}^{\mathrm{NR}(1B)}$ were obtained [36]. From a more realistic nuclear model, we expect to get contributions from both nucleons with smaller contributions from the proton than from neutron operators. For instance, more realistic nuclear models showed that in the context of the minimal SME, the corrections due to the proton operators were only suppressed by a factor of five compared to the neutron operators [29]. Because the comagnetometer experiment is 10^{6}-times more sensitive than the hydrogen experiment, we expect that by using a more realistic nuclear model, we will get better bounds on the proton coefficients from the comagnetometer experiment than from the hydrogen experiment. The best bounds on the proton or neutron coefficients depend on the nuclear models used in the derivation of the Lorentz violation shift. However, in general, the best bounds on the nucleon coefficients will be from hyperfine-Zeeman transitions of the ground states [16,22,35,36], as expected from the discussion in Section 4.

Table 3. Best bounds on the imaginary and real part of the spin-dependent anisotropic nonrelativistic coefficients in the Sun-centered frame for electron, proton, neutron, and muon operators.

Coefficients	Neutron [36] from Xe-He Comagnetometer	Proton and Electron [35] from Hydrogen 1S Splitting	Muon [34] from Muonium 1S Splitting
$H_{w011}^{\mathrm{NR}(0B)}, g_{w011}^{\mathrm{NR}(0B)}$	4×10^{-33} GeV	9×10^{-27} GeV	2×10^{-22} GeV
$H_{w011}^{\mathrm{NR}(1B)}, g_{w011}^{\mathrm{NR}(1B)}$	2×10^{-33} GeV	5×10^{-27} GeV	7×10^{-23} GeV
$H_{w211}^{\mathrm{NR}(0B)}, g_{w211}^{\mathrm{NR}(0B)}$	4×10^{-31} GeV^{-1}	7×10^{-16} GeV^{-1}	1×10^{-11} GeV^{-1}
$H_{w211}^{\mathrm{NR}(1B)}, g_{w211}^{\mathrm{NR}(1B)}$	2×10^{-31} GeV^{-1}	4×10^{-16} GeV^{-1}	6×10^{-12} GeV^{-1}
$H_{w411}^{\mathrm{NR}(0B)}, g_{w411}^{\mathrm{NR}(0B)}$	4×10^{-29} GeV^{-3}	9×10^{-6} GeV^{-3}	2×10^{-1} GeV^{-3}
$H_{w411}^{\mathrm{NR}(1B)}, g_{w411}^{\mathrm{NR}(1B)}$	2×10^{-29} GeV^{-3}	5×10^{-6} GeV^{-3}	8×10^{-2} GeV^{-3}

The energy states involved in the Zeeman transitions used in the He-Xe comagnetometer experiment have total angular momentum quantum number $F = 1/2$, and for that reason, the transitions are only sensitive to nonrelativistic coefficients with $j = 1$. The best bounds on nucleon coefficients with $j > 1$ are from the study of hyperfine-Zeeman transitions of the ground state of cesium using a cesium fountain clock and front the sidereal variations studies with a Ne-Rb-K comagnetometer [36]. In the case of the fountain clock, the energy levels involved in the transition have quantum numbers $F = 3$ and $F = 4$, and these high angular momentum states permit contributions from nonrelativistic coefficients with $j \leq 4$ [36]. The experimental constraints obtained with the atomic fountain clock on sidereal variations are only sensitive to proton coefficients if we assume the nuclear model used in [36]. However, we expect that by using a more realistic nuclear model, we can translate the experimental constraints as bounds on neutron and proton coefficients. Overall, the comagnetometer is more sensitive to smaller frequencies than the cesium atomic clock, and the bounds obtained from the comagnetometer are tighter than the bounds obtained from the atomic fountain clock; however, the atomic fountain clock is sensitive to a greater number of coefficients for Lorentz violation than the comagnetometer.

Hyperfine transitions in large atoms involve nuclear-spin flips, and they are not sensitive to electron Lorentz-violating operators. To study the electron operators, we need to consider hyperfine transitions in light atoms or electron transitions such as optical transitions. The hyperfine-Zeeman transitions of the ground state of hydrogen or the 1 s–2 s transition in hydrogen are only sensitive to electron coefficients with $j \leq 1$. The best bounds on electron coefficients with $j = 2$ are obtained from optical transitions in heavy ions such as $^{40}Ca^{+}$ [24,25,36] and $^{171}Yb^{+}$ [23]. These transitions involve energy levels with high angular momentum. For example, the final energy state for the optical transition in $^{171}Yb^{+}$ has quantum number $F = 3$ [23], and it is sensitive to nonrelativistic coefficients with $j \leq 6$. The final energy state for the optical transition in $^{40}Ca^{+}$ has $F = 5/2$ [24,25], and it is sensitive to coefficients with $j \leq 4$. Overall, the hyperfine-Zeeman transitions of the ground state remain slightly more sensitive to Lorentz violation than the optical transition in $^{171}Yb^{+}$; however, the optical transition is sensitive to a greater number of coefficients for Lorentz violation. Unfortunately, to translate the constraints obtained from the optical transition in $^{171}Yb^{+}$ into bounds on nonminimal coefficients for Lorentz violation, a many-body calculation is needed, and at the moment, this type of calculation has only been done for minimal Lorentz-violating operators [23].

Lorentz violation operators that are isotropic in the laboratory frame cannot be studied using Zeeman-hyperfine transitions of the ground state; see Section 5. The best limits on the coefficients that contribute to f_{sid} in Equation (16) are from an annual variation study of the 1 s–2 s transition in hydrogen [20,35]; see the second and third column of Table 4. As mentioned in Section 5, optical clocks are good candidates to improve these bounds. The best bounds on the isotropic CPT-violating electron and proton terms in the Sun-centered frame are obtained from a comparison of antihydrogen and hydrogen 1 s–2 s transition [36] and for the CPT-even electron term from a comparison between the theoretical and experimental value for the 1 s–2 s transition in positronium [35]. The bounds are shown in Table 5 with the second, fourth, and sixth columns showing the constraints on the electron, proton, and muon isotropic coefficients for Lorentz violation.

Table 4. Best bounds from atomic experiments on effective Cartesian coefficients of mass dimensions $d = 5$ and $d = 6$ in the Sun-centered frame for electron [35], proton [35], and neutron [36] Lorentz-violating operators.

Coefficient	Electron [35] GeV^{4-d}	Proton [35] GeV^{4-d}	Coefficient	Neutron [36] GeV^{4-d}	Coefficient	Neutron [36] GeV^{4-d}
$a_{w_{\text{eff}}}^{(5)TTX}$	$<3.4 \times 10^{-8}$	$<3.4 \times 10^{-8}$	$\tilde{H}_{w_{\text{eff}}}^{(5)X(TXT)}$	$<1 \times 10^{-27}$	$\tilde{g}_{w_{\text{eff}}}^{(6)X(TXTT)}$	$<9 \times 10^{-28}$
$a_{w_{\text{eff}}}^{(5)TTY}$	$<5.6 \times 10^{-8}$	$<5.6 \times 10^{-8}$	$\tilde{H}_{w_{\text{eff}}}^{(5)X(TYT)}$	$<8 \times 10^{-28}$	$\tilde{g}_{w_{\text{eff}}}^{(6)X(TYTT)}$	$<7 \times 10^{-28}$
$a_{w_{\text{eff}}}^{(5)TTY}$	$<1.3 \times 10^{-7}$	$<1.3 \times 10^{-7}$	$\tilde{H}_{w_{\text{eff}}}^{(5)X(TZT)}$	$<2 \times 10^{-27}$	$\tilde{g}_{w_{\text{eff}}}^{(6)X(TZTT)}$	$<2 \times 10^{-27}$
$a_{w_{\text{eff}}}^{(5)KKX}$	$<6.7 \times 10^{-8}$	$<6.7 \times 10^{-8}$	$\tilde{H}_{w_{\text{eff}}}^{(5)Y(TXT)}$	$<8 \times 10^{-28}$	$\tilde{g}_{w_{\text{eff}}}^{(6)Y(TXTT)}$	$<6 \times 10^{-28}$
$a_{w_{\text{eff}}}^{(5)KKY}$	$<1.1 \times 10^{-7}$	$<1.1 \times 10^{-7}$	$\tilde{H}_{w_{\text{eff}}}^{(5)Y(TYT)}$	$<8 \times 10^{-28}$	$\tilde{g}_{w_{\text{eff}}}^{(6)Y(TYTT)}$	$<7 \times 10^{-28}$
$a_{w_{\text{eff}}}^{(5)KKZ}$	$<2.5 \times 10^{-7}$	$<2.5 \times 10^{-7}$	$\tilde{H}_{w_{\text{eff}}}^{(5)Y(TZT)}$	$<2 \times 10^{-27}$	$\tilde{g}_{w_{\text{eff}}}^{(6)Y(TZTT)}$	$<2 \times 10^{-27}$
$c_{w_{\text{eff}}}^{(6)TTTX}$	$<3.3 \times 10^{-5}$	$<1.8 \times 10^{-8}$	$\tilde{H}_{w_{\text{eff}}}^{(5)X(JXJ)}$	$<4 \times 10^{-25}$	$\tilde{g}_{w_{\text{eff}}}^{(6)X(JXJT)}$	$<9 \times 10^{-26}$
$c_{w_{\text{eff}}}^{(6)TTTY}$	$<5.5 \times 10^{-5}$	$<3.0 \times 10^{-8}$	$\tilde{H}_{w_{\text{eff}}}^{(5)X(JYJ)}$	$<3 \times 10^{-25}$	$\tilde{g}_{w_{\text{eff}}}^{(6)X(JYJT)}$	$<7 \times 10^{-26}$
$c_{w_{\text{eff}}}^{(6)TTTZ}$	$<1.3 \times 10^{-4}$	$<6.9 \times 10^{-8}$	$\tilde{H}_{w_{\text{eff}}}^{(5)X(JZJ)}$	$<6 \times 10^{-25}$	$\tilde{g}_{w_{\text{eff}}}^{(6)X(JZJT)}$	$<2 \times 10^{-25}$
$c_{w_{\text{eff}}}^{(6)TKKX}$	$<3.3 \times 10^{-5}$	$<1.8 \times 10^{-8}$	$\tilde{H}_{w_{\text{eff}}}^{(5)Y(JXJ)}$	$<2 \times 10^{-25}$	$\tilde{g}_{w_{\text{eff}}}^{(6)Y(JXJT)}$	$<2 \times 10^{-25}$
$c_{w_{\text{eff}}}^{(6)TKKY}$	$<5.5 \times 10^{-5}$	$<3.0 \times 10^{-8}$	$\tilde{H}_{w_{\text{eff}}}^{(5)Y(JYJ)}$	$<3 \times 10^{-25}$	$\tilde{g}_{w_{\text{eff}}}^{(6)Y(JYJT)}$	$<7 \times 10^{-26}$
$c_{w_{\text{eff}}}^{(6)TKKZ}$	$<1.3 \times 10^{-4}$	$<6.9 \times 10^{-8}$	$\tilde{H}_{w_{\text{eff}}}^{(5)Y(JZJ)}$	$<6 \times 10^{-25}$	$\tilde{g}_{w_{\text{eff}}}^{(6)Y(JZJT)}$	$<2 \times 10^{-25}$
			$\tilde{H}_{w_{\text{eff}}}^{(5)TJTJ}$	$<6 \times 10^{-25}$	$\tilde{g}_{w_{\text{eff}}}^{(6)TJTJT}$	$<5 \times 10^{-25}$

Table 5. Best bounds on the spin-independent isotropic nonrelativistic coefficients in the Sun-centered frame for electron, proton, and muon operators.

	Constraint; Electron		Constraint; Proton		Constraint; Muon
$\lvert a_{e200}^{\text{NR}}\rvert$	$\sim 4 \times 10^{-9}$ GeV^{-1} [36]	$\lvert a_{p200}^{\text{NR}}\rvert$	$\sim 4 \times 10^{-9}$ GeV^{-1} [36]	$\lvert a_{\mu 200}^{\text{NR}}\rvert$	$\sim 3 \times 10^{-5}$ GeV^{-1} [34]
$\lvert c_{e200}^{\text{NR}}\rvert$	$\sim 2 \times 10^{-5}$ GeV^{-1} [35]			$\lvert c_{\mu 200}^{\text{NR}}\rvert$	$\sim 3 \times 10^{-5}$ GeV^{-1} [34]
$\lvert a_{e400}^{\text{NR}}\rvert$	~ 50 GeV^{-3} GeV^{-3} [36]	$\lvert a_{p400}^{\text{NR}}\rvert$	~ 50 GeV^{-3} [36]	$\lvert a_{\mu 400}^{\text{NR}}\rvert$	$\sim 4 \times 10^{5}$ GeV^{-3} [34]
$\lvert c_{e400}^{\text{NR}}\rvert$	$\sim 3 \times 10^{5}$ GeV^{-3} [35]			$\lvert c_{\mu 400}^{\text{NR}}\rvert$	$\sim 4 \times 10^{5}$ GeV^{-3}[34]

Table 4 contains bounds on effective Cartesian coefficients obtained from studying boost effects of the 1 s–2 s transition in hydrogen and the Xe-He comagnetometer. The Lorentz-violating frequency shift in the laboratory frame can be expressed in terms of the nonrelativistic coefficients; however, to consider the annual or sidereal variations due to boost effects, we need to boost the frequency shift from the local laboratory frame to the Sun-centered frame. The nonrelativistic coefficients have simple transformation rules under rotation; however, their transformation under boost transformations is quite complicated, and it is easier to expand the nonrelativistic coefficients in terms of the Cartesian effective coefficients before studying boost effects [35,36]. For that reason, bounds due to boost effects are usually on effective Cartesian coefficients.

Finally, the best bounds on the muon nonrelativistic coefficients were obtained from hyperfine transitions of the ground state of muonium and the 1 s–2 s transition in muonium [17,34,47]; see

Tables 3 and 5. The reader should be aware that many of the bounds reported in [34–36] have not been reproduced in this section. Furthermore, the best bounds on minimal coefficients based on models that do not consider nonminimal terms have also been omitted from the discussion.

9. Outlook

The current bounds on nonminimal Lorentz-violating operators from atomic spectroscopy experiments are based on experimental studies that were designed to impose bounds on the minimal operators [34–36]. Signals associated only with the nonminimal operators such as sidereal variations in higher harmonics of the sidereal frequency have not been constrained experimentally, and they need to be studied to impose bounds on the nonminimal Lorentz-violating operators. For example, time variation studies of the Zeeman-hyperfine transitions of the ground state of cesium only considered the possibility of time variations in the first and second harmonic of the sidereal frequency [22]. The nonminimal Lorentz-violating model predicts that Zeeman-hyperfine transitions are sensitive to time variations up to the fourth harmonic of the sidereal frequency. Experimental constraints on sidereal variations in the third and fourth harmonic of the sidereal frequency will produce limits on coefficients for Lorentz violation that have not been bounded before. The same situation holds for the time variation studies [23–25] of the optical transitions in $^{171}Yb^{+}$ and $^{40}Ca^{+}$. Sidereal variation studies are also needed in the new field of high precision antihydrogen spectroscopy. The antihydrogen collaborations must consider how to implement time variation studies in their experiments if they want to test CPT symmetry systematically [35,36].

The recent publication [36] on the prospects of testing nonminimal Lorentz violation operators in clock comparison experiments used simplistic models for the electron and nuclear configurations. The advantage of using simple models is that they can be easily applied to a large range of systems and the publication intended to recognize the signals for Lorentz violation in a broad range of systems. Using more realistic models will not change the general form of these signals [36]; however, better models are needed to translate experimental constraints on the signals for Lorentz violation into bounds on coefficients. More realistic models have been used in the context of the minimal SME [22–24,29], and similar calculations are needed in the context of the nonminimal SME.

Funding: This work was supported in part by the Department of Energy under grant number DE-SC0010120 and by the Indiana University Center for Spacetime Symmetries.

Acknowledgments: The author would like to thank Jay Tasson for the invitation

Conflicts of Interest: The author declares no conflict of interest.

References

1. Kostelecký, V.A.; Samuel, S. Spontaneous breaking of Lorentz symmetry in string theory. *Phys. Rev. D* **1989**, *39*, 683. [CrossRef]
2. Kostelecký, V.A.; Potting, P. CPT and strings. *Nucl. Phys. B* **1991**, *359*, 545–570. [CrossRef]
3. Mocioiu, I.; Pospelov, M.; Roiban, R. Low energy limits on the antisymmetric tensor field background on the brane and on the non-commutative scale. *Phys. Lett. B* **2000**, *489*, 390–396. [CrossRef]
4. Carroll, S.; Harvey, J.; Kostelecký, V.A.; Lane, C.; Okamoto, T. Noncommutative Field Theory and Lorentz Violation. *Phys. Rev. Lett.* **2001**, *87*, 141601. [CrossRef]
5. Ferrari, A.F.; Gomes, M.; Nascimento, J.R.; Passos, E.; Yu Petrov, A.; da Silva, A.J. Lorentz violation in the linearized gravity. *Phys. Lett. B* **2007**, *652*,174–180. [CrossRef]
6. Gaul, M.; Rovelli, C. Loop quantum gravity and the meaning of diffeomorphism invariance. *Lect. Notes Phys.* **2000**, *541*, 277–324.
7. Bjorken, J.D. Cosmology and the standard model. *Phys. Rev. D* **2003**, *67*, 043508. [CrossRef]
8. Aguilar, P.; Sudarsky, D.; Bonder, Y. Experimental search for a Lorentz invariant spacetime granularity: Possibilities and bounds. *Phys. Rev. D* **2013**, *87*, 064007. [CrossRef]
9. Colladay, D.; Kostelecký, V.A. CPT violation and the standard model. *Phys. Rev. D* **1997**, *55*, 6760–6774. [CrossRef]

10. Colladay, D.; Kostelecký, V.A. Lorentz-violating extension of the standard model. *Phys. Rev. D* **1998**, *58*, 116002. [CrossRef]
11. Bluhm, R.; Kostelecký, V.A.; Russell, N. CPT and Lorentz tests in Penning traps. *Phys. Rev. D* **1998**, *57*, 3932–3943. [CrossRef]
12. Bluhm, R.; Kostelecký, V.A.; Russell, N. CPT and Lorentz Tests in Hydrogen and Antihydrogen. *Phys. Rev. Lett.* **1999**, *82*, 2254–2257. [CrossRef]
13. Kostelecký,V.A.; Lanel, C.D. Constraints on Lorentz violation from clock-comparison experiments. *Phys. Rev. D* **1999**, *60*, 116010. [CrossRef]
14. Bluhm, R.; Kostelecký, V.A.; Lanel, C.D. CPT and Lorentz Tests with Muons. *Phys. Rev. Lett.* **2000**, *84*, 1098–1101. [CrossRef]
15. Bluhm, R.; Kostelecký, V.A.; Lanel, C.D.; Russell, N. Clock-Comparison Tests of Lorentz and CPT Symmetry in Space. *Phys. Rev. Lett.* **2002**, *88*, 090801. [CrossRef]
16. Phillips, D.F.; Humphrey, M.A.; Mattison, E.M.; Stoner, R.E.; Vessot, R.F.C.; Walsworth, R.L. Limit on Lorentz and CPT violation of the proton using a hydrogen maser. *Phys. Rev. D.* **2001**, *63*, 111101. [CrossRef]
17. Hughes, V.W.; Grosse Perdekamp, M.; Kawall, D.; Liu, W.; Jungmann, K.; zu Putlitz, G. Test of CPT and Lorentz Invariance from Muonium Spectroscopy. *Phys. Rev. Lett.* **2001**, *87*, 111804. [CrossRef]
18. Canè, F.; Phillips, D.F.; Rosen,M.S.; Smallwood, C.L.; Stoner, R.E.; Walsworth, R.L.; Kostelecký, V.A. Bound on Lorentz and CPT Violating Boost Effects for the Neutron. *Phys. Rev. Lett.* **2004**, *93*, 230801. [CrossRef]
19. Smiciklas, M.; Brown, J.M.; Cheuk, L.W.; Smullin, S.J.; Romalis, M.V. New Test of Local Lorentz Invariance Using ^{21}Ne-Rb-K Comagnetometer. *Phys. Rev. Lett.* **2011**, *107*, 171604. [CrossRef]
20. Matveev, A.; Parthey, C.G.; Predehl, K.; Alnis, J.; Beyer, A.; Holzwarth, R.; Udem, T.; Wilken, T.; Kolachevsky, N.; Abgrall, M.; et al. Precision Measurement of the Hydrogen 1 s–2 s Frequency via a 920-km Fiber Link. *Phys. Rev. Lett.* **2013**, *110*, 230801. [CrossRef]
21. Allmendinger, F.; Heil, W.; Karpuk, S.; Kilian, W.; Scharth, A.; Schmidt, U.; Schnabel, A.; Sobolev, Y.; Tullney, K. New Limit on Lorentz-Invariance- and CPT-Violating Neutron Spin Interactions Using a Free-Spin-Precession ^{3}He-^{129}Xe Comagnetometer. *Phys. Rev. Lett.* **2014**, *112*, 110801. [CrossRef] [PubMed]
22. Pihan-Le Bars, H.; Guerlin, C.; Lasseri, R.-D.; Ebran, J.-P.; Bailey, Q.G.; Bize, S.; Khan, E.; Wolf, P. Lorentz-symmetry test at Planck-scale suppression with nucleons in a spin-polarized ^{133}Cs cold atom clock. *Phys. Rev. D* **2017**, *95*, 075026. [CrossRef]
23. Sanner, C.; Huntemann, N.; Lange, R.; Tamm, C.; Peik, E.; Safronova, M.S.; Porsev, S.G. Optical clock comparison for Lorentz symmetry testing. *Nature* **2019**, *567*, 204–208. [CrossRef] [PubMed]
24. Pruttivarasin, T.; Ramm, M.; Porsev, S.G.; Tupitsyn, I.I.; Safronova, M.S.; Hohensee, M.A.; Häffner, H. Michelson–Morley analogue for electrons using trapped ions to test Lorentz symmetry. *Nature* **2015**, *517*, 592–595. [CrossRef]
25. Megidish, E.; Broz, J.; Greene, N.; Häffner, H. Improved Test of Local Lorentz Invariance from a Deterministic Preparation of Entangled States. *Phys. Rev. Lett.* **2019**, *122*, 123605. [CrossRef]
26. Flambaum, V.V. ; Romalis, M.V. Limits on Lorentz invariance violation from Coulomb interactions in nuclei and atoms. *Phys. Rev. Lett.* **2017**, *118*, 142501. [CrossRef]
27. Hohensee, M.A.; Leefer, N.; Budker, D.; Harabati, C.; Dzuba, V.A.; Flambaum, V.V. Limits on Violations of Lorentz Symmetry and the Einstein Equivalence Principle using Radio-Frequency Spectroscopy of Atomic Dysprosium. *Phys. Rev. Lett.* **2013**, *111*, 050401. [CrossRef]
28. Harabati, C.; Dzuba, V.A.; Flambaum, V.V.; Hohensee, M.A. Effects of Lorentz-symmetry violation on the spectra of rare-earth ions in a crystal field. *Phys. Rev. A* **2015**, *92*, 040101. [CrossRef]
29. Stadnik, Y.V.; Flambaum,V.V. Nuclear spin-dependent interactions: searches for WIMP, axion and topological defect dark matter, and tests of fundamental symmetries. *Eur. Phys. J. C* **2015**, *75*, 110. [CrossRef]
30. Kostelecký, V.A.; Mewes, M. Electrodynamics with Lorentz-violating operators of arbitrary dimension. *Phys. Rev. D* **2009**, *80*, 015020. [CrossRef]
31. Kostelecký, V.A.; Mewes, M. Neutrinos with Lorentz-violating operators of arbitrary dimension. *Phys. Rev. D* **2012**, *85*, 096005. [CrossRef]
32. Kostelecký, V.A.; Mewes, M. Fermions with Lorentz-violating operators of arbitrary dimension. *Phys. Rev. D* **2013**, *88*, 96006. [CrossRef]
33. Kostelecký, V.A.; Li, Z. Gauge field theories with Lorentz-violating operators of arbitrary dimension. *Phys. Rev. D* **2019**, *99*, 056016. [CrossRef]

34. Gomes, A.H.; Kostelecký, V.A.; Vargas, A.J. Laboratory tests of Lorentz and CPT symmetry with muons. *Phys. Rev. D* **2014**, *90*, 076009. [CrossRef]
35. Kostelecký, V.A.; Vargas, A.J. Lorentz and CPT tests with hydrogen, antihydrogen, and related systems. *Phys. Rev. D* **2015**, *92*, 056002. [CrossRef]
36. Kostelecký, V.A.; Vargas, A.J. Lorentz and CPT Tests with Clock-Comparison Experiments. *Phys. Rev. D* **2018**, *98*, 036003. [CrossRef]
37. Ding, Y.; Kostelecký, V.A. Lorentz-violating spinor electrodynamics and Penning traps. *Phys. Rev. D* **2016**, *94*, 056008. 10.1103/PhysRevD.94.056008. [CrossRef]
38. Yoder, T.J.; Adkins, G.S. Higher order corrections to the hydrogen spectrum from the standard-model extension. *Phys. Rev. D* **2012**, *86*, 116005. [CrossRef]
39. Foldy, L.L.; Wouthuysen, S.A. On the Dirac Theory of Spin 1/2 Particles and its Non-Relativistic Limit. *Phys. Rev.* **1950**, *78*, 29–36. [CrossRef]
40. Kostelecký, V.A.; Mewes, M. Signals for Lorentz violation in electrodynamics. *Phys. Rev. D* **2002**, *66*, 056005. [CrossRef]
41. Ahmadi, M.; Alves, B.X.R.; Baker, C.J.; Bertsche, W.; Butler, E.; Capra, A.; Carruth, C.; Cesar, C.L.; Charlton, M.; Cohen, S.; et al. Observation of the hyperfine spectrum of antihydrogen. *Nature* **2017**, *548*, 66–69. [CrossRef] [PubMed]
42. Malbrunot, C.; Amsler, C.; Arguedas Cuendis, S.; Breuker, H.; Dupre, P.; Fleck, M.; Higaki, H.; Kanai, Y.; Kolbinger, B.; Kuroda, N.; et al. The ASACUSA antihydrogen and hydrogen program: Results and prospects. *Philos. Trans. R. Soc. A* **2018**, *376*, 20170273. [CrossRef] [PubMed]
43. Cacciapuoti, L.; Salomon, C. Space clocks and fundamental tests: The ACES experiment. *Eur. Phys. J. Spec. Top.* **2009**, *172*, 57–68. [CrossRef]
44. Greenberg, O.W. CPT Violation Implies Violation of Lorentz Invariance. *Phys. Rev. Lett.* **2002**, *86*, 231602. [CrossRef]
45. Parthey, C.G.; Matveev, A.; Alnis, J.; Bernhardt, B.; Beyer, A.; Holzwarth, R.; Maistrou, A.; Pohl, R.; Predehl, K.; Udem, T.; et al. Improved Measurement of the Hydrogen 1 s–2 s Transition Frequency. *Phys. Rev. Lett.* **2011**, *107*, 203001. [CrossRef]
46. Ahmadi, M.; Alves, B.X.R.; Baker, C.J.; Bertsche, W.; Butler, E.; Capra, A.; Carruth, C.; Cesar, C.L.; Charlton, M.; Cohen, S.; et al. Characterization of the 1S–2S transition in antihydrogen. *Nature* **2018**, *557*, 71–75. [CrossRef]
47. Strasser, P.; Abe, M.; Aoki, M.; Choi, S.; Fukao, Y.; Higashi, Y.; Higuchi, T.; Iinuma, H.; Ikedo, Y.; Ishida, K.; et al. New precise measurements of muonium hyperfine structure at J-PARC MUSE. *EPJ Web Conf.* **2019**, *198*, 00003. . [CrossRef]

symmetry

MDPI

Article

Lorentz and CPT Tests Using Penning Traps

Yunhua Ding

Department of Physics, Gettysburg College, Gettysburg, PA 17325, USA; yding@gettysburg.edu

Received: 30 August 2019; Accepted: 25 September 2019; Published: 1 October 2019

Abstract: The theoretical prospects for quantum electrodynamics with Lorentz-violating operators of mass dimensions up to six are revisited in this work. The dominant effects due to Lorentz and CPT violation are studied in measurements of magnetic moments of particles confined in Penning traps. Using recently reported experimental results, new coefficients for Lorentz violation are constrained and existing bounds of various coefficients are improved.

Keywords: lorentz violation; CPT violation; penning trap

1. Introduction

The recent measurements of the proton and antiproton magnetic moments have reached record sensitivities of ppb levels by confining the particles in electromagnetic fields using a Penning trap [1,2]. For the electron magnetic moment, a similar Penning-trap experiment has also been carried out in an impressive precision of ppt level [3]. Experiments measuring the positron magnetic moment are currently under development to aim for a comparable precision as that of the electron [4,5]. These highly precise measurements in Penning-trap experiments offer a great way to study fundamental symmetries, including Lorentz and CPT invariances, the foundation of Einstein's theory of relativity. It has been shown that tiny deviations from relativity could naturally emerge in a fundamental theory unifying gravity with quantum physics at the Plank scale $M_P \sim 10^{19}$ GeV, such as strings [6,7]. In recent years, many high-precision tests of relativity in various subfields of physics have been performed to search for Lorentz- and CPT-violating signals [8], including the spectroscopic studies of particles confined in Penning traps.

Any tiny violation effects arising from a large unknown energy scale are well described by effective field theory. The comprehensive framework describing Lorentz violation in the context of effective field theory is given by the Standard-Model Extension (SME) [9–11], which is constructed by adding all possible Lorentz-violating terms into the action of General Relativity and the Standard Model. Each of the terms is formed from a coordinate-independent contraction of a Lorentz-violating operator with the corresponding controlling coefficient. In the context of effective field theory, CPT violation is accompanied by the breaking of Lorentz symmetry [9,10,12], so the SME also describes general CPT-violating effects. The SME provides a general framework to study possible effects due to Lorentz and CPT violation and the parameters in any specific model characterizing these violations can be matched to a suitable subset of the SME coefficients.

The minimal SME contains Lorentz-violating operators of mass dimensions up to four, which is power-counting renormalizable in Minkowski spacetime. Lorentz-violating operators of larger mass dimensions can be viewed as corrections at higher orders to the low-energy limit. Study of the nonminimal SME is of interest in many different contexts of physics, such as the causality and stability [13,14], the pseudo-Riemann–Finsler geometry [15–18], the mixing of Lorentz-violating operators of different mass dimensions [19], Lorentz-violating models in supersymmetry [20], and noncommutative Lorentz-violating quantum electrodynamics [21–23].

In a Penning-trap experiment, the measurable effects due to Lorentz and CPT violation given by the minimal SME include the charge-to-mass ratio and the magnetic moment of the confined particle [24,25], through the changes in the anomaly and cyclotron frequencies. The published work on studying the minimal-SME effects involves comparison of the anomaly frequencies of the electron and positron [26], sidereal-variation analysis of the electron anomaly frequency [27,28], and measurements of cyclotron frequencies of the H^- ion relative to that of the antiproton [29,30].

In the nonminimal SME, additional Lorentz- and CPT-violating effects beyond the minimal SME can be generated by the interaction between the confined particle and the electromagnetic fields in the trap. The general theory of quantum electrodynamics with Lorentz- and CPT-violating operators of mass dimensions up to six has been constructed in Ref. [31]. Recently, this treatment was generalized to include operators of arbitrary mass dimension using gauge field theory [32]. In this work we focus on further studies of Lorentz- and CPT-violating effects in the nonminimal sector of the SME by using the recent Penning-trap results, which include the sidereal-variation analysis of the anomaly frequencies for electrons [28] and the comparison of magnetic moments for both protons and antiprotons [1,2], to obtain new and improved constraints on the SME coefficients. The results from this work are complementary to existing studies of Penning-trap experiments [31], the muon anomalous magnetic moment [33,34], clock-frequency comparison [35], and spectroscopy of hydrogen, antihydrogen, and other related systems [36].

This work is organized as follows. In Section 2, we review the theory of quantum electrodynamics with Lorentz- and CPT-violating operators of mass dimensions up to six. We use perturbation theory to obtain the dominant shifts arising from Lorentz violation to the energy levels of the trapped fermion, and then derive the contributions to the cyclotron and anomaly frequencies. The discussion of coordinate transformation is given at the end of this section. We next turn in Section 3 the experimental applications related to Penning traps and use the reported results to extract new limits on various SME coefficients, including some that were not constrained in the literature. The constraints on the SME coefficients obtained in this work are summarized in Section 4.

2. Theory

The theoretical prospects of Lorentz- and CPT-violating quantum electrodynamics in Penning-trap experiments have been studied in Ref. [31]. In this section, we review the main results.

For a Dirac fermion field ψ of charge q and mass m_ψ confined in an external electromagnetic field specified by potential A_μ, the conventional gauge-invariant Lagrange density $\mathcal{L}_0$ takes the form

$$\mathcal{L}_0 = \tfrac{1}{2}\overline{\psi}(\gamma^\mu iD_\mu - m_\psi)\psi + \text{h.c.}, \tag{1}$$

where the covariant derivative is given by the minimal coupling $iD_\mu = (i\partial_\mu - qA_\mu)$ and h.c. means Hermitian conjugate. The general Lorentz-violating Lagrange density that preserves U(1) gauge invariance for the Dirac fermion field ψ can be constructed by adding contraction terms of Lorentz-violating operators with the corresponding SME coefficients [9,10],

$$\mathcal{L}_\psi = \tfrac{1}{2}\overline{\psi}(\gamma^\mu iD_\mu - m_\psi + \widehat{\mathcal{Q}})\psi + \text{h.c.}, \tag{2}$$

where $\widehat{\mathcal{Q}}$ is a general 4×4 spinor matrix involving the covariant derivative iD_μ and the electromagnetic field tensor $F_{\alpha\beta} \equiv \partial_\alpha A_\beta - \partial_\beta A_\alpha$. The hermiticity of the Lagrange density (2) guarantees that $\widehat{\mathcal{Q}}$ satisfies condition $\widehat{\mathcal{Q}} = \gamma_0\widehat{\mathcal{Q}}^\dagger\gamma_0$. In the limit of the free Dirac fermion with $A_\alpha = 0$, Ref. [37] has studied the propagation of the fermion field ψ at arbitrary mass dimension. A similar analysis of the quadratic terms in the photon sector at arbitrary mass dimension has been presented in Ref. [38], as well as extensions to other sectors, e.g., nonminimal neutrino [39] and gravity [40].

In this work we focus on the dominant Lorentz-violating effects including the photon-fermion interaction beyond the minimal SME and restrict our attention to operators in the Lagrange density (2) with mass dimensions up to six. The related full Lagrange density (2) can then be expressed as two parts, the conventional Lagrange density $\mathcal{L}_0$ plus a series of contributions $\mathcal{L}^{(d)}$ according to the mass dimension of the operators, presented in Ref. [31]. We note that the nonminimal operators in the Lagrange density (2) generate a new type of SME coefficients with subscript F which control the sizes of interactions involving the fermion spinors ψ and the electromagnetic field strength $F_{\alpha\beta}$. For example, the dimension-five terms in the Lagrange density (2) contain a contribution involving $b_F^{(5)\mu\alpha\beta} F_{\alpha\beta}\overline{\psi}\gamma_5\gamma_\mu\psi$.

In the Lagrange density (2), the presence of Lorentz-violating operators modifies the conventional Dirac equation for a fermion in electromagnetic fields and generates corrections $\delta\mathcal{H}$ to the Hamiltonian. Since no Lorentz violation has been observed so far, any corrections must be tiny. We thus can treat $\delta\mathcal{H}$ as a perturbative contribution and apply perturbation theory to obtain the dominant Lorentz- and CPT-violating shifts in energy levels,

$$\delta E_{n,s} = \langle \chi_{n,s} | \delta\mathcal{H} | \chi_{n,s} \rangle, \tag{3}$$

where $E_{n,s}$ are unperturbed eigenstates of nth level and s is the spin state taking values of $+1$ and -1 for spin up and down, respectively.

From the modified Dirac equation given by the Lagrange density (2)

$$(p \cdot \gamma - m + \widehat{\mathcal{Q}})\psi = 0, \tag{4}$$

the exact Hamiltonian $\mathcal{H}$ can be defined as

$$\mathcal{H}\psi \equiv p^0\psi = \gamma_0(\boldsymbol{p} \cdot \boldsymbol{\gamma} + m - \widehat{\mathcal{Q}})\psi, \tag{5}$$

where p^0 is the exact energy. The exact perturbative Hamiltonian $\delta\mathcal{H}$ can then be identified as

$$\delta\mathcal{H} = -\gamma_0\widehat{\mathcal{Q}}. \tag{6}$$

It is challenging to construct $\delta\mathcal{H}$ directly as terms proportional to higher powers of momentum appear in $\widehat{\mathcal{Q}}$ and these terms in general contain the perturbative Hamiltonian $\mathcal{H}$ itself. However, any contributions to $\delta\mathcal{H}$ due to the exact Hamiltonian $\mathcal{H}$ are at second or higher orders in the SME coefficients. To obtain the leading-order corrections, one thus can evaluate p^0 in $\widehat{\mathcal{Q}}$ at the unperturbed eigenstates $E_{n,s}$,

$$\delta\mathcal{H} \approx -\gamma_0\widehat{\mathcal{Q}}|_{p^0 \to E_{n,s}}. \tag{7}$$

In a Penning-trap experiment the primary observables of interest are transition frequencies generated by the energy shifts due to the electromagnetic fields in the trap. Among the key frequencies are the Larmor frequency for spin precession $\nu_L \equiv \omega_L/2\pi$ and the cyclotron frequency $\nu_c \equiv \omega_c/2\pi$. The difference of the two frequencies gives the anomaly frequency $\nu_L - \nu_c = \nu_a \equiv \omega_a/2\pi$ [41]. The measurements of the magnetic moment and the related g factor of a particle confined in the trap can then be determined by the following ratio,

$$\frac{\nu_L}{\nu_c} \equiv \frac{\omega_L}{\omega_c} = \frac{g}{2}. \tag{8}$$

The above frequencies can be shifted in the presence of Lorentz and CPT violation, as the energies are modified by Equation (3). To show the explicit results of the shifts, we choose the apparatus frame with cartesian coordinates $x^a \equiv (x^1, x^2, x^3)$ so that the magnetic field $\boldsymbol{B} = B\hat{x}_3$ points at the positive x^3 direction and fix the electromagnetic potential gauge to be $A^\mu = (0, x_2 B, 0, 0)$. For a confined particle of fermion-flavor $w = e, p$ for electrons and protons, and of charge polarity $\sigma = +1, -1$ for carrying positive and negative charges, there is no leading-order contribution from Lorentz and CPT violation to the cyclotron frequencies,

$$\delta\omega_c^w = \delta E_{1,\sigma}^w - \delta E_{0,\sigma}^w \approx 0. \tag{9}$$

The dominant Lorentz- and CPT-violating contributions appear in the shifts to the anomaly frequencies,

$$\delta\omega_a^w = \delta E_{0,-\sigma}^w - \delta E_{1,\sigma}^w = 2\widetilde{b}_w^3 - 2\widetilde{b}_{F,w}^{33} B, \tag{10}$$

where the tilde quantities are defined by

$$\begin{aligned} \widetilde{b}_w^3 &= b_w^3 + H_w^{12} - m_w d_w^{30} - m_w g_w^{120} + m_w^2 b_w^{(5)300} + m_w^2 H_w^{(5)1200} - m_w^3 d_w^{(6)3000} - m_w^3 g_w^{(6)12000}, \\ \widetilde{b}_{F,w}^{33} &= b_{F,w}^{(5)312} + H_{F,w}^{(5)1212} - m_w d_{F,w}^{(6)3012} - m_w g_{F,w}^{(6)12012}. \end{aligned} \tag{11}$$

Here the superscripts (d) of the nonminimal SME coefficients in the tilde quantities (11) denote the mass dimensions of the corresponding coefficients.

For the Lorentz- and CPT-violating shifts to the cyclotron and anomaly frequencies of the corresponding antifermion of flavor $\overline{w}$, a similar analysis can be carried out by reversing the signs of the CPT-odd SME coefficients in Equations (9) and (10). As with the fermion case, the leading-order contributions to the cyclotron frequencies vanish,

$$\delta\omega_c^{\overline{w}} = \delta E_{1,\sigma}^{\overline{w}} - \delta E_{0,\sigma}^{\overline{w}} \approx 0, \tag{12}$$

and the shifts to the anomaly frequencies are given by

$$\delta\omega_a^{\overline{w}} = \delta E_{0,-\sigma}^{\overline{w}} - \delta E_{1,\sigma}^{\overline{w}} = -2\widetilde{b}_w^{*3} + 2\widetilde{b}_{F,w}^{*33} B, \tag{13}$$

where the two sets of starred tilde coefficients are defined as

$$\begin{aligned} \widetilde{b}_w^{*3} &= b_w^3 - H_w^{12} + m_w d_w^{30} - m_w g_w^{120} + m_w^2 b_w^{(5)300} - m_w^2 H_w^{(5)1200} + m_w^3 d_w^{(6)3000} - m_w^3 g_w^{(6)12000}, \\ \widetilde{b}_{F,w}^{*33} &= b_{F,w}^{(5)312} - H_{F,w}^{(5)1212} + m_w d_{F,w}^{(6)3012} - m_w g_{F,w}^{(6)12012}. \end{aligned} \tag{14}$$

The index pair 12 in the tilde quantities (11) and (14) is antisymmetric and transforms under rotation like a single 3 index, thus the shifts (10) and (13) in the anomaly frequencies for both fermions and antifermions depend on only the $\hat{x}_3$ direction, as expected from the cylindrical symmetry of the trap.

The above results (10) and (13) show that the dominant contributions in the anomaly frequencies for a trapped fermion and antifermion of flavor w in a Penning trap are given by the four tilde combinations $\widetilde{b}_w^3$, $\widetilde{b}_{F,w}^{33}$, $\widetilde{b}_w^{*3}$, and $\widetilde{b}_{F,w}^{*33}$. The results are valid in the apparatus frame, in which the magnetic field is aligned with the positive $\hat{x}_3$ axis. However, this apparatus frame is noninertial due to the Earth's rotation. The standard canonical frame adopted in the literature to compare results from different experiments searching for Lorentz violation is the Sun-centered frame [42,43], with the cartesian coordinates $X^J \equiv (X, Y, Z)$. In this frame, the Z axis is aligned with the rotation axis of the Earth and the X axis points towards the vernal equinox in the year 2000. The Sun-centered frame is approximately inertial in a typical time scale for an experiment. To relate the SME coefficients from the Sun-centered frame to the apparatus frame, we introduce a third frame called the standard laboratory frame with cartesian coordinates $x^j \equiv (x, y, z)$. The z axis in this frame points towards the local zenith and the x axis is aligned with the local south. The choice of the positive $\hat{x}_3$ axis in the apparatus frame to be aligned with the direction of the magnetic field may result in a nonzero angle to the $\hat{z}$ axis, so the transformation $x^a = R^{aj} x^j$ relating (x, y, z) in the standard laboratory frame to (x^1, x^2, x^3) in the apparatus frame involves a rotation matrix R^{aj} specified in general by suitable Euler angles α, β, and γ,

$$R^{aj} = \begin{pmatrix} \cos\gamma & \sin\gamma & 0 \\ -\sin\gamma & \cos\gamma & 0 \\ 0 & 0 & 1 \end{pmatrix} \begin{pmatrix} \cos\beta & 0 & -\sin\beta \\ 0 & 1 & 0 \\ \sin\beta & 0 & \cos\beta \end{pmatrix} \begin{pmatrix} \cos\alpha & \sin\alpha & 0 \\ -\sin\alpha & \cos\alpha & 0 \\ 0 & 0 & 1 \end{pmatrix}. \tag{15}$$

Neglecting boost effects, which are at the order of 10^{-4}, the relationship $x^j = R^{jJ}x^J$ between (X, Y, Z) in the Sun-centered frame and (x, y, z) in the standard laboratory frame can be obtained by applying the following rotation matrix [42,43]

$$R^{jJ}(T_\oplus) = \begin{pmatrix} \cos\chi\cos\omega_\oplus T_\oplus & \cos\chi\sin\omega_\oplus T_\oplus & -\sin\chi \\ -\sin\omega_\oplus T_\oplus & \cos\omega_\oplus T_\oplus & 0 \\ \sin\chi\cos\omega_\oplus T_\oplus & \sin\chi\sin\omega_\oplus T_\oplus & \cos\chi \end{pmatrix}, \qquad (16)$$

where $\omega_\oplus \simeq 2\pi/(23\text{ h }56\text{ min})$ is the sidereal frequency of the Earth's rotation, $T_\oplus$ is the local sidereal time, and the angle χ specifies the laboratory colatitude.

To relate the time T in the Sun-centered frame to the time t in the standard laboratory frame, it is often convenient to match the origin of t with the local sidereal time $T_\oplus$, by defining its origin at the moment when the y axis in the standard laboratory frame lies along the Y axis in the Sun-centered frame. For a laboratory with longitude λ in units of degrees, this choice offsets t from T by an integer number of the Earth's sidereal rotations plus an additional shift

$$T_0 \equiv T - T_\oplus \simeq \frac{(66.25^\circ - \lambda)}{360^\circ}(23.934\text{ hr}). \qquad (17)$$

The above discussion shows that the relationship between (X, Y, Z) in the Sun-centered frame and (x^1, x^2, x^3) in the apparatus frame is given by

$$x^a(T_\oplus) = R^{aj}R^{jJ}(T_\oplus)X^J. \qquad (18)$$

The transformation (18) generates the dependence on the sidereal time of the SME coefficients observed in the apparatus frame. To show the explicit dependence of the shifts to the anomaly frequencies (10), consider a fermion of flavor w confined in a Penning trap with the magnetic field aligned with the local zenith and located at colatitude χ, applying the transformation matrix (16) yields the results

$$\widetilde{b}^3_w = \widetilde{b}^Z_w\cos\chi + (\widetilde{b}^X_w\cos\omega_\oplus T_\oplus + \widetilde{b}^Y_w\sin\omega_\oplus T_\oplus)\sin\chi, \qquad (19)$$

and

$$\begin{aligned}\widetilde{b}^{33}_{F,w} = \widetilde{b}^{ZZ}_{F,w} + \tfrac{1}{2}(\widetilde{b}^{XX}_{F,w} + \widetilde{b}^{YY}_{F,w} - 2\widetilde{b}^{ZZ}_{F,w})\sin^2\chi + (\widetilde{b}^{(XZ)}_{F,w}\cos\omega_\oplus T_\oplus + \widetilde{b}^{(YZ)}_{F,w}\sin\omega_\oplus T_\oplus)\sin 2\chi \\ +\Big(\tfrac{1}{2}(\widetilde{b}^{XX}_{F,w} - \widetilde{b}^{YY}_{F,w})\cos 2\omega_\oplus T_\oplus + \widetilde{b}^{(XY)}_{F,w}\sin 2\omega_\oplus T_\oplus\Big)\sin^2\chi,\end{aligned} \qquad (20)$$

where the parenthesis around two indices (JK) in the tilde coefficients means symmetrization and is defined as $(JK) = (JK + KJ)/2$. Similar results for the shifts to the anomaly frequencies (13) of antifermions can also be derived by substituting the tilde coefficients with the starred tilde coefficients. In more general cases with the magnetic field pointing a generic direction, information on the Euler angles α, β, γ in Equation (15) are needed to obtain the explicit results.

The results given above show that the physical observables in a Penning-trap experiment involving electrons, positrons, protons, and antiprotons are the 36 independent tilde quantities $\widetilde{b}^J_w$, $\widetilde{b}^{*J}_w$, $\widetilde{b}^{(JK)}_{F,w}$, and $\widetilde{b}^{*(JK)}_{F,w}$ in the Sun-centered frame. Performing a sidereal-variation analysis of the anomaly frequencies can give access to 28 of the coefficients as the other 8 contributions proportional to $\widetilde{b}^Z_w$, $\widetilde{b}^{*Z}_w$, $\widetilde{b}^{ZZ}_{F,w}$, and $\widetilde{b}^{*(ZZ)}_{F,w}$ are independent of sidereal time. A comparison of the results from two different Penning-trap experiments is therefore required to study these 8 combinations of coefficients for Lorentz violation that produce constant shifts to the anomaly frequencies.

3. Experiment

3.1. Harvard Experiment

The recent measurement of the electron g factor performed at Harvard University has reached a precision of 0.28 ppt [3]. A sidereal-variation analysis of the anomaly frequencies for the electron was performed to search for variations in the sidereal time of the Earth's rotation [28]. The data was analyzed for oscillations over time and was fit by a five-parameter sinusoid model at the sidereal frequencies of $\omega_\oplus$ and $2\omega_\oplus$, yielding a 2σ limit in the amplitudes of the harmonic oscillation of $|\delta\nu_a^e| \lesssim 0.05$ Hz. This result corresponds to $|\delta\omega_a^e| \lesssim 2 \times 10^{-25}$ GeV in natural units with $c = \hbar = 1$. The magnetic field adopted in the experiment is $B = 5.36$ T in the local upward direction and the geometrical colatitude of this experiment is $\chi = 47.6°$. Taking one sidereal oscillation at a time places bounds

$$\left(\left(\widetilde{b}_e^X - (2\times 10^{-15}\,\mathrm{GeV}^2)\widetilde{b}_{F,e}^{(XZ)}\right)^2 + \left(\widetilde{b}_e^Y - (2\times 10^{-15}\,\mathrm{GeV}^2)\widetilde{b}_{F,e}^{(YZ)}\right)^2\right)^{1/2} \lesssim 2\times 10^{-25}\,\mathrm{GeV} \quad (21)$$

in the first harmonic and

$$\left(\left(10^{-15}\,\mathrm{GeV}^2(\widetilde{b}_{F,e}^{XX} - \widetilde{b}_{F,e}^{YY})\right)^2 + \left(10^{-15}\,\mathrm{GeV}^2\widetilde{b}_{F,e}^{(XY)}\right)^2\right)^{1/2} \lesssim 2\times 10^{-25}\,\mathrm{GeV} \quad (22)$$

in the second harmonic, respectively. The above results not only lead to a factor of four improvement compared to the existing constraints obtained by a similar analysis of the Penning-trap experiment searching for first-harmonic variation at the University of Washington [27,31], but also produce the first-time bounds on tilde coefficients $\widetilde{b}_{F,e}^{(XX)} - \widetilde{b}_{F,e}^{(YY)}$ and $\widetilde{b}_{F,e}^{(XY)}$ as they only appear in the second harmonic of the sidereal oscillation.

The experiments to measure the magnetic moment of a trapped positron are currently under development at Harvard University and Northwestern University [4,5]. Performing a similar sidereal-variation analysis of the anomaly frequency would offer not only the first-time limits on the starred tilde coefficients $\widetilde{b}_e^{*J}$, $\widetilde{b}_{F,e}^{*(JK)}$, but would also constrain the CPT-odd coefficients in Equations (11) and (14) by comparing with measurements of the electron. The constant parts in the sidereal variations of the tilde coefficients $\widetilde{b}_e^J$, $\widetilde{b}_{F,e}^{(JK)}$, $\widetilde{b}_e^{*J}$, and $\widetilde{b}_{F,e}^{*(JK)}$ could also be studied by this comparison.

3.2. BASE Experiments at Mainz and CERN

The BASE collaboration has recently measured the proton magnetic moment at a record sensitivity of 0.3 ppb using a Penning trap located at Mainz [1], improving their previous best result [44] by a factor of 11. A precision of 1.5 ppb of the antiproton magnetic moment measurement has also been achieved by the same group using a similar Penning trap located at CERN [2]. A study of sidereal variations of the anomaly frequencies for both protons and antiprotons is currently being performed at BASE and this could, in principle, provide sensitivities to various tilde coefficients $\widetilde{b}_p^J$, $\widetilde{b}_{F,p}^{(JK)}$, $\widetilde{b}_p^{*J}$ and $\widetilde{b}_{F,p}^{*(JK)}$. Another version of this experiment is planned to be performed at CERN by the BASE collaboration to measure the magnetic moments for both protons and antiprotons using quantum logic readout [45], which will allow rapid measurements of the anomaly frequencies for the proton and antiproton. This would offer an excellent opportunity to conduct the sidereal-variation analysis, as well as to constrain the constant parts in the harmonics of the above coefficients through a direct comparison of the two measurements.

Here we combine the published results from the two recent BASE experiments [1,2] to obtain constraints on the SME coefficients in the Sun-centered frame. A comparison between the two measured g factors for protons and antiprotons gives

$$\frac{g_p}{2} - \frac{g_{\overline{p}}}{2} = \frac{\omega_a^p}{\omega_c^p} - \frac{\omega_a^{\overline{p}}}{\omega_c^{\overline{p}}} = \frac{2}{\omega_c^p \omega_c^{\overline{p}}} \left(\Sigma\omega_c^p \Delta\omega_a^p - \Delta\omega_c^p \Sigma\omega_a^p \right), \tag{23}$$

where the differences and sums of the cyclotron and anomaly frequencies are defined as

$$\begin{aligned} \Delta\omega_c^p &\equiv \tfrac{1}{2}(\omega_c^p - \omega_c^{\overline{p}}), \\ \Sigma\omega_c^p &\equiv \tfrac{1}{2}(\omega_c^p + \omega_c^{\overline{p}}), \\ \Delta\omega_a^p &\equiv \tfrac{1}{2}(\delta\omega_a^p - \delta\omega_a^{\overline{p}}), \\ \Sigma\omega_a^p &\equiv \tfrac{1}{2}(\delta\omega_a^p + \delta\omega_a^{\overline{p}}). \end{aligned} \tag{24}$$

For the proton magnetic moment measured at Mainz, the experiment is located at $\chi \simeq 40.0^\circ$ and the applied magnetic field $B \simeq 1.9$ T points $\theta = 18^\circ$ from local south in the counterclockwise direction, generating a cyclotron frequency $\omega_c^p = 2\pi \times 28.96$ MHz [1]. For the antiproton magnetic moment measurement at CERN, the trap is located at $\chi^* \simeq 43.8^\circ$ and the magnetic field $B^* \simeq 1.95$ T points $\theta^* = 120^\circ$ from local south in the counterclockwise direction, producing a different cyclotron frequency $\omega_c^{\overline{p}} = 2\pi \times 29.66$ MHz [2]. Since the measurements of the frequencies for both experiments were performed over an extended time period, any sidereal variations could be plausibly assumed to be averaged out, leaving only the constant parts in the tilde coefficients. Therefore, applying the general transformation (18) together with the related experimental quantities yields the following expressions for the time-independent parts in $\Delta\omega_a^p$ and $\Sigma\omega_a^p$,

$$\begin{aligned} \Delta\omega_a^p &= \widetilde{b}_p^3 - \widetilde{b}_{F,p}^{33} B + \widetilde{b}_p^{*3} - \widetilde{b}_{F,p}^{*33} B^* \\ &= -\widetilde{b}_p^Z \cos\theta \sin\chi - \widetilde{b}_p^{*Z} \cos\theta^* \sin\chi^* \\ &\quad -\tfrac{1}{2}(\widetilde{b}_{F,p}^{XX} + \widetilde{b}_{F,p}^{YY}) B (\cos^2\theta \cos^2\chi + \sin^2\theta) - \tfrac{1}{2}(\widetilde{b}_{F,p}^{*XX} + \widetilde{b}_{F,p}^{*YY}) B^* (\cos^2\theta^* \cos^2\chi^* + \sin^2\theta^*) \\ &\quad -\widetilde{b}_{F,p}^{ZZ} B \cos^2\theta \sin^2\chi - \widetilde{b}_{F,p}^{*ZZ} B^* \cos^2\theta^* \sin^2\chi^*, \\ \Sigma\omega_a^p &= \widetilde{b}_p^3 - \widetilde{b}_{F,p}^{33} B - \widetilde{b}_p^{*3} + \widetilde{b}_{F,p}^{*33} B^* \\ &= -\widetilde{b}_p^Z \cos\theta \sin\chi + \widetilde{b}_p^{*Z} \cos\theta^* \sin\chi^* \\ &\quad -\tfrac{1}{2}(\widetilde{b}_{F,p}^{XX} + \widetilde{b}_{F,p}^{YY}) B (\cos^2\theta \cos^2\chi + \sin^2\theta) + \tfrac{1}{2}(\widetilde{b}_{F,p}^{*XX} + \widetilde{b}_{F,p}^{*YY}) B^* (\cos^2\theta^* \cos^2\chi^* + \sin^2\theta^*) \\ &\quad -\widetilde{b}_{F,p}^{ZZ} B \cos^2\theta \sin^2\chi + \widetilde{b}_{F,p}^{*ZZ} B^* \cos^2\theta^* \sin^2\chi^*. \end{aligned} \tag{25}$$

Substituting expressions (25) into the difference (23) and adopting the numerical values of the experimental quantities given above, the reported results for the measurements of g factors from both BASE experiments give the following 2σ limit

$$\begin{aligned} &\Big| \widetilde{b}_p^Z - 0.6\widetilde{b}_p^{*Z} + (2\times10^{-16}\,\mathrm{GeV}^2)(\widetilde{b}_{F,p}^{XX} + \widetilde{b}_{F,p}^{YY}) + (2\times10^{-16}\,\mathrm{GeV}^2)\widetilde{b}_{F,p}^{ZZ} \\ &\quad + (2\times10^{-16}\,\mathrm{GeV}^2)(\widetilde{b}_{F,p}^{*XX} + \widetilde{b}_{F,p}^{*YY}) + (7\times10^{-17}\,\mathrm{GeV}^2)\widetilde{b}_{F,p}^{*ZZ} \Big| \lesssim 8\times10^{-25}\,\mathrm{GeV}. \end{aligned} \tag{26}$$

4. Sensitivity

To get some intuition for the scope of the constraints (21), (22), and (26), a common practice is to assume only one individual tilde coefficient is nonzero at a time. Considering no Lorentz and CPT violation has been observed so far, this procedure offers a reasonable measure of the estimated limits on each tilde coefficient by ignoring any cancellations among them. We list in Table 1 the resulting constraints on the tilde coefficients from this work and include the previous limits obtained in Ref. [31], as well as recent improved results presented in Ref. [2], for a direct comparison. In the electron sector, Table 1 shows that not only a factor of four improvement for the limits on the tilde coefficients $\widetilde{b}_e^X$, $\widetilde{b}_e^Y$, $\widetilde{b}_e^{(XZ)}$, and $\widetilde{b}_e^{(YZ)}$ has been achieved, but also that new coefficients $\widetilde{b}_{F,e}^{(XY)}$ and $\widetilde{b}_{F,e}^{*XX} - \widetilde{b}_{F,e}^{*YY}$ have been constrained. In the proton sector, the limits on the tilde coefficients have been improved by factors of up to three compared to the existing results [2]. The constraints on the tilde coefficients that are not sensitive to the corresponding work are left blank in Table 1. Please note that only 18 out of the 36 coefficients for Lorentz violation related to Penning-trap experiments have been constrained so far. A sidereal-variation analysis for the measurements of the magnetic moments of protons and antiprotons would permit access to other various components of the tilde coefficients in the proton sector.

Table 1. New and improved constraints on the SME coefficients.

Coefficient	Previous Constraint in [31]	Recent Result in [2]	This Work
$\lvert\widetilde{b}_e^X\rvert$	$< 6 \times 10^{-25}$ GeV		$< 1 \times 10^{-25}$ GeV
$\lvert\widetilde{b}_e^Y\rvert$	$< 6 \times 10^{-25}$ GeV		$< 1 \times 10^{-25}$ GeV
$\lvert\widetilde{b}_e^Z\rvert$	$< 7 \times 10^{-24}$ GeV		
$\lvert\widetilde{b}_e^{*Z}\rvert$	$< 7 \times 10^{-24}$ GeV		
$\lvert\widetilde{b}_{F,e}^{XX} + \widetilde{b}_{F,e}^{YY}\rvert$	$< 2 \times 10^{-8}$ GeV^{-1}		
$\lvert\widetilde{b}_{F,e}^{ZZ}\rvert$	$< 8 \times 10^{-9}$ GeV^{-1}		
$\lvert\widetilde{b}_{F,e}^{(XY)}\rvert$			$< 2 \times 10^{-10}$ GeV^{-1}
$\lvert\widetilde{b}_{F,e}^{(XZ)}\rvert$	$< 4 \times 10^{-10}$ GeV^{-1}		$< 1 \times 10^{-10}$ GeV^{-1}
$\lvert\widetilde{b}_{F,e}^{(YZ)}\rvert$	$< 4 \times 10^{-10}$ GeV^{-1}		$< 1 \times 10^{-10}$ GeV^{-1}
$\lvert\widetilde{b}_{F,e}^{*XX} + \widetilde{b}_{F,e}^{*YY}\rvert$	$< 2 \times 10^{-8}$ GeV^{-1}		
$\lvert\widetilde{b}_{F,e}^{*XX} - \widetilde{b}_{F,e}^{*YY}\rvert$			$< 4 \times 10^{-10}$ GeV^{-1}
$\lvert\widetilde{b}_{F,e}^{*ZZ}\rvert$	$< 8 \times 10^{-9}$ GeV^{-1}		
$\lvert\widetilde{b}_p^Z\rvert$	$< 2 \times 10^{-21}$ GeV	$< 1.8 \times 10^{-24}$ GeV	$< 8 \times 10^{-25}$ GeV
$\lvert\widetilde{b}_p^{*Z}\rvert$	$< 6 \times 10^{-21}$ GeV	$< 3.5 \times 10^{-24}$ GeV	$< 1 \times 10^{-24}$ GeV
$\lvert\widetilde{b}_{F,p}^{XX} + \widetilde{b}_{F,p}^{YY}\rvert$	$< 1 \times 10^{-5}$ GeV^{-1}	$< 1.1 \times 10^{-8}$ GeV^{-1}	$< 4 \times 10^{-9}$ GeV^{-1}
$\lvert\widetilde{b}_{F,p}^{ZZ}\rvert$	$< 1 \times 10^{-5}$ GeV^{-1}	$< 7.8 \times 10^{-9}$ GeV^{-1}	$< 3 \times 10^{-9}$ GeV^{-1}
$\lvert\widetilde{b}_{F,p}^{*XX} + \widetilde{b}_{F,p}^{*YY}\rvert$	$< 2 \times 10^{-5}$ GeV^{-1}	$< 7.4 \times 10^{-9}$ GeV^{-1}	$< 3 \times 10^{-9}$ GeV^{-1}
$\lvert\widetilde{b}_{F,p}^{*ZZ}\rvert$	$< 8 \times 10^{-6}$ GeV^{-1}	$< 2.7 \times 10^{-8}$ GeV^{-1}	$< 1 \times 10^{-8}$ GeV^{-1}

5. Summary

In conclusion, we present in this work the general theory for quantum electrodynamics with Lorentz- and CPT-violating operators of mass dimensions up to six and study the dominant effects arising from Lorentz and CPT violation in Penning-trap experiments involving confined particles. Recently reported results of magnetic moments of the confined particles are used to improve existing bounds on various SME coefficients, and to constrain new coefficients as well. The results obtained in this work are summarized in Table 1. The methodology we outline in this work using Equation (23) to derive these constraints can be used as a generic way to study Lorentz and CPT violation involving comparisons of results from different Penning-trap experiments. The high sensitivities of the measurements in current and forthcoming experiments offer strong motivation to continue the efforts of studying Lorentz and CPT violation with great potential to uncover any possible tiny signals.

Funding: This research was funded by the Department of Energy grant number DE-SC0010120 and by the Indiana University Center for Spacetime Symmetries.

Acknowledgments: The author would like to thank Jay Tasson for the invitation and V. Alan Kostelecký for useful discussion.

Conflicts of Interest: The author declares no conflict of interest.

References

1. Schneider, G.; Mooser, A.; Bohman, M.; Schön, N.; Harrington, J.; Higuchi, T.; Nagahama, H.; Sellner, S.; Smorra, C.; Blaum, K.; et al. Double-trap measurement of the proton magnetic moment at 0.3 parts per billion precision. *Science* **2017**, *358*, 1081. [CrossRef] [PubMed]
2. Smorra, C.; Sellner, S.; Borchert, M.J.; Harrington, J.A.; Higuchi, T.; Nagahama, H.; Tanaka, T.; Mooser, A.; Schneider, G.; Bohman, M.; et al. A parts-per-billion measurement of the antiproton magnetic moment. *Nature* **2017**, *550*, 371. [CrossRef] [PubMed]
3. Hanneke, D.; Hoogerheide, S.F.; Gabrielse, G. Cavity control of a single-electron quantum cyclotron: Measuring the electron magnetic moment. *Phys. Rev. A* **2011**, *83*, 052122. [CrossRef]
4. Hoogerheide, S.F.; Dorr, J.C.; Novitski, E.; Gabrielse, G. High efficiency positron accumulation for high-precision magnetic moment experiments. *Rev. Sci. Instrum.* **2015**, *86*, 053301. [CrossRef] [PubMed]
5. Gabrielse, G.; Fayer, S.E.; Myers, T.G.; Fan, X. Towards an improved test of the Standard Model's most precise prediction. *Atoms* **2019**, *7*, 45. [CrossRef]
6. Kostelecký, V.A.; Samuel, S. Spontaneous breaking of Lorentz symmetry in string theory. *Phys. Rev. D* **1989**, *39*, 683. [CrossRef] [PubMed]
7. Kostelecký, V.A.; Potting, R. CPT and strings. *Nucl. Phys. B* **1991** *359*, 545. [CrossRef]
8. Kostelecký, V.A.; Russell, N. Data tables for Lorentz and CPT violation. *Rev. Mod. Phys.* **2011**, *83*, 11. [CrossRef]
9. Colladay, D.; Kostelecký, V.A. CPT violation and the Standard Model. *Phys. Rev. D* **1997**, *55*, 6760. [CrossRef]
10. Colladay, D.; Kostelecký, V.A. Lorentz-violating extension of the Standard Model. *Phys. Rev. D* **1998**, *58*, 116002. [CrossRef]
11. Kostelecký, V.A. Gravity, Lorentz violation, and the Standard Model. *Phys. Rev. D* **2004**, *69*, 105009. [CrossRef]
12. Greenberg, O.W. CPT violation implies violation of Lorentz invariance. *Phys. Rev. Lett.* **2002**, *89*, 231602. [CrossRef] [PubMed]
13. Kostelecký, V.A.; Lehnert, R. Stability, causality, and Lorentz and CPT violation. *Phys. Rev. D* **2001**, *63*, 065008. [CrossRef]
14. Drummond, I.T. Quantum field theory in a multimetric background. *Phys. Rev. D* **2013**, *88*, 025009. [CrossRef]
15. Kostelecký, V.A. Riemann-Finsler geometry and Lorentz-violating kinematics. *Phys. Lett. B* **2011**, *701*, 137. [CrossRef]
16. Silva, J.E.G.; Maluf, R.V.; Almeida, C.A.S. A nonlinear dynamics for the scalar field in Randers spacetime. *Phys. Lett. B* **2017**, *766*, 263. [CrossRef]
17. Foster, J.; Lehnert, R. Classical-physics applications for Finsler *b* space. *Phys. Lett. B* **2015**, *746*, 164. [CrossRef]
18. Edwards, B.; Kostelecký, V.A. Riemann-Finsler geometry and Lorentz-violating scalar fields. *Phys. Lett. B* **2018**, *786*, 319. [CrossRef]
19. Cambiaso, M.; Lehnert, R.; Potting, R. Asymptotic states and renormalization in Lorentz-violating quantum field theory. *Phys. Rev. D* **2014**, *90*, 065003. [CrossRef]
20. Belich, H.; Bernald, L.D.; Gaete, P.; Helayël-Neto, J.A.; Leal, F.J.L. Aspects of CPT-even Lorentz-symmetry violating physics in a supersymmetric scenario. *Eur. Phys. J. C* **2015**, *75*, 291. [CrossRef]
21. Carroll, S.M.; Harvey, J.A.; Kostelecký, V.A.; Lane, C.D.; Okamoto, T. Noncommutative field theory and Lorentz violation. *Phys. Rev. Lett.* **2001**, *87*, 141601. [CrossRef] [PubMed]
22. Chaichian, M.; Sheikh-Jabbari, M.M.; Tureanu, A. Non-commutativity of space-time and the hydrogen atom spectrum. *Eur. Phys. J. C* **2004**, *36*, 251. [CrossRef]
23. Hayakawa, M. Perturbative analysis on infrared aspects of noncommutative QED on R^4. *Phys. Lett. B* **2000**, *478*, 394. [CrossRef]
24. Bluhm, R.; Kostelecký, V.A.; Russell, N. Testing CPT with anomalous magnetic moments. *Phys. Rev. Lett.* **1997**, *79*, 1432. [CrossRef]
25. Bluhm, R.; Kostelecký, V.A.; Russell, N. CPT and Lorentz tests in Penning traps. *Phys. Rev. D* **1998**, *57*, 3932. [CrossRef]
26. Dehmelt, H.G.; Mittleman, R.K.; Van Dyck R.S.; Schwinberg, P. Past electron-positron $g-2$ experiments yielded sharpest bound on CPT violation for point particles. *Phys. Rev. Lett.* **1999**, *83*, 4694. [CrossRef]

27. Mittleman, R.K.; Ioannou, I.I.; Dehmelt, H.G.; Russell, N. Bound on CPT and Lorentz symmetry with a trapped electron. *Phys. Rev. Lett.* **1999**, *83*, 2116. [CrossRef]
28. Hanneke, D. Cavity Control in a Single-Electron Quantum Cyclotron: An Improved Measurement of the Electron Magnetic Moment. Ph.D. Thesis, Harvard University, Cambridge, MA, USA, 2007.
29. Gabrielse, G.; Khabbaz, A.; Hall, D.S.; Heimann, C.; Kalinowsky, H.; Jhe, W. Precision mass spectroscopy of the antiproton and proton using simultaneously trapped particles. *Phys. Rev. Lett.* **1999**, *82*, 3198. [CrossRef]
30. Ulmer, S.; Smorra, C.; Mooser, A.; Franke, K.; Nagahama, H.; Schneider, G.; Higuchi, T.; Van Gorp, S.; Blaum, K.; Matsuda, Y.; et al. High-precision comparison of the antiproton-to-proton charge-to-mass ratio. *Nature* **2015**, *524*, 196. [CrossRef]
31. Ding, Y.; Kostelecký, V.A. Lorentz-violating spinor electrodynamics and Penning traps. *Phys. Rev. D* **2016**, *94*, 056008. [CrossRef]
32. Li, Z.; Kostelecký, V.A. Gauge field theories with Lorentz-violating operators of arbitrary dimension. *Phys. Rev. D* **2019**, *99*, 056016.
33. Bluhm, R.; Kostelecký, V.A.; Lane, C.D. CPT and Lorentz tests with muons. *Phys. Rev. Lett.* **2000**, *84*, 1098. [CrossRef] [PubMed]
34. Gomes, A.H.; Kostelecký, V.A.; Vargas, A. Laboratory tests of Lorentz and C P T symmetry with muons. *Phys. Rev. D* **2014**, *90*, 076009. [CrossRef]
35. Kostelecký, V.A.; Vargas, A. Lorentz and CPT Tests with Clock-Comparison Experiments. *Phys. Rev. D* **2018**, *98*, 036003. [CrossRef]
36. Kostelecký, V.A.; Vargas, A. Lorentz and CPT tests in hydrogen, antihydrogen, and related systems. *Phys. Rev. D* **2015**, *92*, 056002. [CrossRef]
37. Kostelecký, V.A.; Mewes, M. Fermions with Lorentz-violating operators of arbitrary dimension. *Phys. Rev. D* **2013**, *88*, 096006. [CrossRef]
38. Kostelecký, V.A.; Mewes, M. Electrodynamics with Lorentz-violating operators of arbitrary dimension. *Phys. Rev. D* **2009**, *80*, 015020. [CrossRef]
39. Kostelecký, V.A.; Mewes, M. Neutrinos with Lorentz-violating operators of arbitrary dimension. *Phys. Rev. D* **2012**, *85*, 096005. [CrossRef]
40. Kostelecký, V.A.; Mewes, M. Testing local Lorentz invariance with gravitational waves. *Phys. Lett. B* **2016**, *757*, 510. [CrossRef]
41. Brown, L.S.; Gabrielse, G. Geonium theory: Physics of a single electron or ion in a Penning trap. *Rev. Mod. Phys.* **1986**, *58*, 233. [CrossRef]
42. Kostelecký, V.A.; Lane, C.D. Constraints on Lorentz Violation from Clock-Comparison Experiments. *Phys. Rev. D* **1999**, *60*, 116010. [CrossRef]
43. Kostelecký, V.A.; Mewes, M. Signals for Lorentz violation in electrodynamics. *Phys. Rev. D* **2002**, *66*, 056005. [CrossRef]
44. Mooser, A.; Ulmer, S.; Blaum, K.; Franke, K.; Kracke, H.; Leiteritz, C.; Quint, W.; Rodegheri, C.C.; Smorra, C.; Walz, J. Direct high-precision measurement of the magnetic moment of the proton. *Nature* **2014**, *509*, 596. [CrossRef] [PubMed]
45. Meiners, T.; Niemann, M.; Paschke, A.G.; Borchert, M.; Idel, A.; Mielke, J.; Voges, K.; Bautista-Salvador, A.; Lehnert, R.; Ulmer, S.; et al. Towards sympathetic laser cooling and detection of single (anti-)protons. In *Proceedings of the Seventh Meeting on CPT and Lorentz Symmetry*; Kostelecký, V.A., Ed.; World Scientific: Singapore, 2017.

symmetry

MDPI

Article

Formal Developments for Lorentz-Violating Dirac Fermions and Neutrinos

João Alfíeres Andrade de Simões dos Reis [1,2] and Marco Schreck [1,*]

1 Departamento de Física, Universidade Federal do Maranhão, Campus Universitário do Bacanga, São Luís–MA 65080-805, Brazil; jalfieres@gmail.com

2 Departamento de Física, Centro de Educação, Ciências Exatas e Naturais, Universidade Estadual do Maranhão, Cidade Universitária Paulo VI, São Luís–MA 65055-310, Brazil

* Correspondence: marco.schreck@ufma.br; Tel.: +55-(98)3272-8293

Received: 31 August 2019; Accepted: 17 September 2019; Published: 24 September 2019

Abstract: The current paper is a technical work that is focused on Lorentz violation for Dirac fermions as well as neutrinos, described within the nonminimal Standard-Model Extension. We intend to derive two theoretical results. The first is the full propagator of the single-fermion Dirac theory modified by Lorentz violation. The second is the dispersion equation for a theory of N neutrino flavors that enables the description of both Dirac and Majorana neutrinos. As the matrix structure of the neutrino field operator is very involved for generic N, we will use sophisticated methods of linear algebra to achieve our objectives. Our main finding is that the neutrino dispersion equation has the same structure in terms of Lorentz-violating operators as that of a modified single-fermion Dirac theory. The results will be valuable for phenomenological studies of Lorentz-violating Dirac fermions and neutrinos.

Keywords: Lorentz and *CPT* violation; Standard-Model Extension; Dirac fermions; Dirac neutrinos; Majorana neutrinos; determinants of block matrices

---

1. Introduction

Neutrinos are both interesting and elusive particles. According to the Standard Model of particle physics, there are three neutrino flavors that correspond to the flavors of the three charged leptons—the electron neutrino ν_e, the muon neutrino ν_μ and the tau neutrino ν_τ. Neutrinos do not carry electric charge and are only produced in processes mediated by the weak interaction [1]. When neutrinos propagate a distance, the probability of detecting a certain flavor changes with time. These neutrino oscillations are quantum mechanical in nature. They have their origin in the fact that the eigenstates of the kinematic Hamiltonian and the flavor eigenstates produced in interactions do not correspond to each other. Instead, these two distinct bases are related by the unitary PMNS matrix [2]. Neutrino oscillations indicate that neutrinos have mass, although their mass is that tiny to not have been measured directly, so far. Therefore, they practically propagate with the speed of light.

Measuring the *CP*-violating phase that is contained in the PMNS matrix is currently one of the hot topics in neutrino physics. Apart from *CP*-violation, neutrinos may be subject to *CPT* violation due to physics at the Planck scale such as strings [3,4]. Since they travel almost with the speed of light, they are strongly boosted with respect to the Sun-centered equatorial frame [5] and *CPT*-violating effects may accumulate over long distances. A violation of *CPT* symmetry implies a violation of Lorentz invariance in the context of effective field theory [6], whereby it makes sense to use neutrinos as a testbed for the search for Lorentz violation. To do so, a general comprehensive framework to parameterize Lorentz violation is desirable. The latter is provided by the Standard-Model Extension (SME). Its minimal version including field operators of mass dimensions 3 and 4 was developed in References [7,8]. The nonminimal SME, which involves field operators of arbitrary mass dimensions, was constructed in a series of papers [9–11] for photons, neutrinos and single Dirac fermions. Within this framework, each Lorentz-violating contribution is a proper contraction of a field operator and a background field. The latter are composed of preferred spacetime directions and controlling coefficients parameterizing the strength of Lorentz violation.

Phenomenology in the SME neutrino sector has been performed in a large collection of papers [12–33], where this list is not claimed to be complete. All constraints obtained for Lorentz violation in the neutrino sector are compiled in the data tables [5]. Interestingly, it was also observed that neutrino oscillations could be explained by Lorentz-violating *massless* neutrinos in certain models.

The current article must be considered as a technical work that can be the preparing base for further forthcoming phenomenological investigations. We have two objectives. The first is to present the general result for the propagator of a single Dirac fermion in the nonminimal SME. The second is to obtain the dispersion equation of the nonminimal neutrino sector. The latter goal will be accomplished with a powerful method to compute determinants of large matrices that decompose into smaller blocks.

A description of Lorentz violation in the neutrino sector based on the SME permits N neutrino flavors and allows for both Dirac and Majorana neutrinos. Majorana neutrinos are characterized by the property of being identical to their own antiparticles such that neutrino-antineutrino mixing can occur. Therefore, the differential operator that appears in the field equations of the theory is not simply a (4×4) matrix in spinor space such as for a single Dirac fermion. For a set of N neutrinos of either Dirac or Majorana type, it is a $(8N \times 8N)$ matrix instead. The determinant of this matrix directly corresponds to the modified dispersion equation for neutrinos. Even for 3 flavors, its computation is cumbersome.

In principle, it is possible to consider an observer frame with only a single nonzero coefficient and to compute the determinant by brute force with computer algebra. Such a direct approach has several disadvantages, though. First, an observer frame with a single nonzero controlling coefficient is a very special case. Second, the result of the determinant is most probably still messy and its structure is supposedly not very illuminating. Therefore, it would be desirable to employ a technique that allows for a covariant and general treatment of the problem. Although the result is still expected to be complicated due to the high dimensionality of the matrix, the method to be used can be applied to obtain the dispersion equation for an arbitrary number of flavors. Thus, it may be of interest for someone who wants to include sterile neutrinos in their analysis, which are beyond the scope of the SME.

The paper is organized as follows. Section 2 gives a brief introduction to the SME fermion sector. The properties most important to us are discussed and several definitions are introduced. In Section 3 we derive the full propagator of the modified Dirac fermion sector. In Section 4 the very base of the SME neutrino sector is described, as well as the algorithm that we intend to use to obtain the neutrino dispersion equation. The individual steps of the calculation are carried out and explained, too. We state the central result in Section 5 and discuss it subsequently, whereby some properties of the first-order dispersion relations are obtained in Section 6. A brief comment on classical Lagrangians in the neutrino sector follows in Section 7. Finally, all findings are summarized and concluded on in Section 8. Relations and definitions that are not of primary interest to the reader are relegated to Appendixes A and B. Natural units with the conventions $\hbar = c = 1$ will be used unless otherwise stated.

2. Basic Properties of the SME Fermion Sector

The fermion sector of the SME describes a single Dirac fermion subject to Lorentz-violating background fields that permit a construction of an observer Lorentz-invariant Lagrange density. The minimal fermion sector was introduced in Reference [8] and its properties were investigated in Reference [34]. The minimal framework was complemented by the nonminimal contributions in Reference [11]. Our analysis will be carried out within the nonminimal SME, whereby we take over the notation of the latter reference and also mainly refer to formulas stated in that paper.

The framework rests on the Lagrange density of Equations (1) and (2) in [11]. The corresponding modified Dirac equation reads

$$\mathcal{D}\psi = 0\,, \quad \mathcal{D} = \not{p} - m_\psi \mathbb{1}_4 + \hat{\mathcal{Q}}\,, \tag{1a}$$

$$\hat{\mathcal{Q}} = \hat{\mathcal{S}}\mathbb{1}_4 + \mathrm{i}\hat{\mathcal{P}}\gamma^5 + \hat{\mathcal{V}}^\mu\gamma_\mu + \hat{\mathcal{A}}^\mu\gamma^5\gamma_\mu + \frac{1}{2}\hat{\mathcal{T}}^{\mu\nu}\sigma_{\mu\nu}\,. \tag{1b}$$

In the latter, ψ is a spinor, m_ψ the fermion mass, and γ^μ denote the usual Dirac matrices that satisfy the Clifford algebra $\{\gamma^\mu, \gamma^\nu\} = 2\eta^{\mu\nu}\mathbb{1}_4$ with the Minkowski metric $\eta_{\mu\nu}$ of signature $(+,-,-,-)$. Furthermore, $\mathbb{1}_n$ is the n-dimensional identity matrix, $\gamma_5 = \gamma^5 \equiv \mathrm{i}\gamma^0\gamma^1\gamma^2\gamma^3$ is the chiral Dirac matrix and $\sigma_{\mu\nu} \equiv (\mathrm{i}/2)[\gamma_\mu, \gamma_\nu]$ involves the commutator of two Dirac matrices. The Lorentz-violating operator $\hat{\mathcal{Q}}$ is decomposed in terms of the 16 matrices $\{\Gamma^A\} \equiv \{\mathbb{1}_4, \gamma^5, \gamma^\mu, \mathrm{i}\gamma^5\gamma^\mu, \sigma^{\mu\nu}\}$. This set forms a basis of (4×4) matrices and the dual basis $\{\Gamma_A\}$ is obtained by lowering the Lorentz indices with the Minkowski metric. The basis obeys an orthogonality relation of the form $\mathrm{Tr}(\Gamma_A\Gamma^B) = 4\delta_A{}^B$ where Tr denotes the trace in spinor space. Lorentz violation is contained in a scalar $\hat{\mathcal{S}}$, a vector $\hat{\mathcal{V}}$, and a two-tensor operator $\hat{\mathcal{T}}$. Additionally, a pseudo-scalar $\hat{\mathcal{P}}$ and a pseudo-vector $\hat{\mathcal{A}}$ occur when the behavior of the operators under parity transformations is taken into account. Tensors of higher rank than these do not exist, as more complicated matrices in spinor space can always be mapped in some way to the 16 matrices mentioned before by using identities such as those listed in Appendix A.

It is worth pointing out the structure of the operators contained in $\hat{\mathcal{Q}}$. They are constructed as sums of operators of increasing mass dimension suitably contracted with controlling coefficients. In momentum space they can be written as the following infinite sums:

$$\hat{\mathcal{S}} = \sum_{d=3}^{\infty} \mathcal{S}^{(d)\alpha_1\alpha_2\ldots\alpha_{d-3}} p_{\alpha_1} p_{\alpha_2} \ldots p_{\alpha_{d-3}}\,, \tag{2a}$$

$$\hat{\mathcal{P}} = \sum_{d=3}^{\infty} \mathcal{P}^{(d)\alpha_1\alpha_2\ldots\alpha_{d-3}} p_{\alpha_1} p_{\alpha_2} \ldots p_{\alpha_{d-3}}\,, \tag{2b}$$

$$\hat{\mathcal{V}}^\mu = \sum_{d=3}^{\infty} \mathcal{V}^{(d)\mu\alpha_1\alpha_2\ldots\alpha_{d-3}} p_{\alpha_1} p_{\alpha_2} \ldots p_{\alpha_{d-3}}\,, \tag{2c}$$

$$\hat{\mathcal{A}}^\mu = \sum_{d=3}^{\infty} \mathcal{A}^{(d)\mu\alpha_1\alpha_2\ldots\alpha_{d-3}} p_{\alpha_1} p_{\alpha_2} \ldots p_{\alpha_{d-3}}\,, \tag{2d}$$

$$\hat{\mathcal{T}}^{\mu\nu} = \sum_{d=3}^{\infty} \mathcal{T}^{(d)\mu\nu\alpha_1\alpha_2\ldots\alpha_{d-3}} p_{\alpha_1} p_{\alpha_2} \ldots p_{\alpha_{d-3}}\,, \tag{2e}$$

where d is the mass dimension of the field operator that a specific controlling coefficient such as $\mathcal{S}^{(d)\alpha_1\ldots\alpha_{d-3}}$ is contracted with. These decompositions can be extracted from Equation (3) in [11]. Each controlling coefficient has a mass dimension of $4-d$ and is independent of the spacetime coordinates to preserve energy and momentum.

Evaluating the determinant of the Dirac operator $\mathcal{D}$ leads to the dispersion equation for a single fermion. The latter is given by Equation (39) of [11]:

$$\Delta = 0\,, \quad \Delta = (\hat{\mathcal{S}}_-^2 - \hat{\mathcal{T}}_-^2)(\hat{\mathcal{S}}_+^2 - \hat{\mathcal{T}}_+^2) + \hat{\mathcal{V}}_-^2\hat{\mathcal{V}}_+^2 - 2\hat{\mathcal{V}}_-\cdot(\hat{\mathcal{S}}_-\eta + 2\mathrm{i}\hat{\mathcal{T}}_-)\cdot(\hat{\mathcal{S}}_+\eta - 2\mathrm{i}\hat{\mathcal{T}}_+)\cdot\hat{\mathcal{V}}_+\,, \tag{3a}$$

with the definitions

$$\hat{\mathcal{S}}_\pm \equiv -m_\psi + \hat{\mathcal{S}} \pm \mathrm{i}\hat{\mathcal{P}}\,, \quad \hat{\mathcal{V}}_\pm^\mu \equiv p^\mu + \hat{\mathcal{V}}^\mu \pm \hat{\mathcal{A}}^\mu\,, \quad \hat{\mathcal{T}}_\pm^{\mu\nu} \equiv \frac{1}{2}(\hat{\mathcal{T}}^{\mu\nu} \pm \mathrm{i}\tilde{\hat{\mathcal{T}}}^{\mu\nu})\,, \tag{3b}$$

and the dual of the two-tensor operator:

$$\widetilde{\hat{\mathcal{T}}}^{\mu\nu} \equiv \frac{1}{2}\varepsilon^{\mu\nu\varrho\sigma}\hat{\mathcal{T}}_{\varrho\sigma}\,. \tag{3c}$$

The latter contains the totally antisymmetric Levi-Civita symbol $\varepsilon^{\mu\nu\varrho\sigma}$ based on the convention $\varepsilon^{0123} = 1$. In the remainder of the paper, a definition of the observer scalars

$$\hat{X} \equiv \frac{1}{4}\hat{\mathcal{T}}_{\mu\nu}\hat{\mathcal{T}}^{\mu\nu}\,,\quad \hat{Y} \equiv \frac{1}{4}\hat{\mathcal{T}}_{\mu\nu}\widetilde{\hat{\mathcal{T}}}^{\mu\nu}\,, \tag{4}$$

turns out to be fruitful. It is also beneficial to define the following combination of operators that we will make frequent use of:

$$\widetilde{\hat{\mathcal{T}}}^{\mu\nu}_{\mathrm{gen}} \equiv \widetilde{\hat{\mathcal{T}}}^{\mu\nu} - \frac{1}{\hat{\mathcal{S}} - m_{\psi}}\left[(p+\hat{\mathcal{V}})^{\mu}\hat{\mathcal{A}}^{\nu} - \hat{\mathcal{A}}^{\mu}(p+\hat{\mathcal{V}})^{\nu}\right]\,,\quad \hat{\mathcal{T}}^{\mu\nu}_{\mathrm{gen}} \equiv \frac{1}{2}\varepsilon^{\mu\nu\varrho\sigma}(\widetilde{\hat{\mathcal{T}}}_{\mathrm{gen}})_{\varrho\sigma}\,. \tag{5}$$

Note that the operator previously introduced reduces to the effective dual two-tensor operator in Equation (25) of Reference [11] at first order in Lorentz violation, which is why we denote it by the index "gen" standing for "generalized."

3. General Modified Dirac Propagator

The investigations to be performed in the neutrino sector require an evaluation of the single-fermion propagator (Green's function in momentum space) S. This result has not been obtained so far for the full nonminimal fermion sector, which is why we would like to state it here. The propagator is directly connected to the inverse of the Dirac operator in momentum space: $\mathcal{D}S = S\mathcal{D} = \mathbb{1}_4$. It must be possible to express the propagator in terms of the basis $\{\Gamma^A\}$ mentioned before. Therefore, it can be written in the form

$$\mathrm{i}S = \frac{\mathrm{i}}{\Delta}\left(\hat{\mathcal{S}}^{(p)}\mathbb{1}_4 + \mathrm{i}\hat{\mathcal{P}}^{(p)}\gamma^5 + \hat{\mathcal{V}}^{(p)\mu}\gamma_{\mu} + \hat{\mathcal{A}}^{(p)\mu}\gamma^5\gamma_{\mu} + \frac{1}{2}\hat{\mathcal{T}}^{(p)\mu\nu}\sigma_{\mu\nu}\right)\,, \tag{6}$$

where $\Delta = \det(\mathcal{D})$ and the index (p) of each individual contribution stands for "propagator." Note that we introduced a prefactor of i to follow the conventions of Reference [35]. The denominator Δ corresponds to the left-hand side of Equation (3). The individual contributions can be obtained by multiplying the inverse with each of the 16 Dirac matrices and computing the trace of the matrix product. In addition, we make use of the orthogonality relation for these matrices, which leads to:

$$\hat{\mathcal{S}}^{(p)} = \frac{\Delta}{4}\mathrm{Tr}(\mathbb{1}_4\mathcal{D}^{-1})\,,\quad \hat{\mathcal{P}}^{(p)} = -\mathrm{i}\frac{\Delta}{4}\mathrm{Tr}(\gamma^5\mathcal{D}^{-1})\,,\quad \hat{\mathcal{V}}^{(p)\mu} = \frac{\Delta}{4}\mathrm{Tr}(\gamma^{\mu}\mathcal{D}^{-1})\,, \tag{7a}$$

$$\hat{\mathcal{A}}^{(p)\mu} = -\frac{\Delta}{4}\mathrm{Tr}(\gamma^5\gamma^{\mu}\mathcal{D}^{-1})\,,\quad \hat{\mathcal{T}}^{(p)\mu\nu} = \frac{\Delta}{4}\mathrm{Tr}(\sigma^{\mu\nu}\mathcal{D}^{-1})\,. \tag{7b}$$

Now, these contributions are explicitly given by

$$\hat{\mathcal{S}}^{(p)} = -(\hat{\mathcal{S}} - m_{\psi})(2\Theta - \hat{\mathcal{T}}_{\mu\nu}\hat{\mathcal{T}}^{\mu\nu}_{\mathrm{gen}}) - 2\hat{Y}\hat{\mathcal{P}}\,, \tag{8a}$$

$$\hat{\mathcal{P}}^{(p)} = -(\hat{\mathcal{S}} - m_{\psi})(2\hat{Y} - \hat{\mathcal{T}}_{\mu\nu}\widetilde{\hat{\mathcal{T}}}^{\mu\nu}_{\mathrm{gen}}) + 2\Theta\hat{\mathcal{P}}\,, \tag{8b}$$

$$\hat{\mathcal{V}}^{(p)\mu} = 2\left[\Theta(p+\hat{\mathcal{V}})^{\mu} - \hat{\mathcal{T}}^{\mu\nu}\hat{\mathcal{T}}_{\nu\varrho}(p+\hat{\mathcal{V}})^{\varrho} - (\hat{\mathcal{S}} - m_{\psi})\widetilde{\hat{\mathcal{T}}}^{\mu\nu}_{\mathrm{gen}}\hat{\mathcal{A}}_{\nu} + \hat{\mathcal{T}}^{\mu\nu}\hat{\mathcal{A}}_{\nu}\hat{\mathcal{P}}\right]\,, \tag{8c}$$

$$\hat{\mathcal{A}}^{(p)\mu} = 2\left[\Theta\hat{\mathcal{A}}^{\mu} - \hat{\mathcal{T}}^{\mu\nu}\hat{\mathcal{T}}_{\nu\varrho}\hat{\mathcal{A}}^{\varrho} - (\hat{\mathcal{S}} - m_{\psi})\widetilde{\hat{\mathcal{T}}}^{\mu\nu}_{\mathrm{gen}}(p+\hat{\mathcal{V}})_{\nu} + \hat{\mathcal{T}}^{\mu\nu}(p+\hat{\mathcal{V}})_{\nu}\hat{\mathcal{P}}\right]\,, \tag{8d}$$

$$\hat{\mathcal{T}}^{(p)\mu\nu} = 2\Big[(\hat{\mathcal{S}} - m_\psi)^2\hat{\mathcal{T}}^{\mu\nu}_{\text{gen}} - \Theta\hat{\mathcal{T}}^{\mu\nu} + \hat{\Upsilon}\widetilde{\hat{\mathcal{T}}}^{\mu\nu} + \big[\hat{\mathcal{T}}^{\mu\varrho}(p+\hat{\mathcal{V}})^\nu - (p+\hat{\mathcal{V}})^\mu\hat{\mathcal{T}}^{\nu\varrho}\big](p+\hat{\mathcal{V}})_\varrho \\ -(\hat{\mathcal{T}}^{\mu\varrho}\hat{\mathcal{A}}^\nu - \hat{\mathcal{A}}^\mu\hat{\mathcal{T}}^{\nu\varrho})\hat{\mathcal{A}}_\varrho - (\hat{\mathcal{S}} - m_\psi)\hat{\mathcal{P}}\widetilde{\hat{\mathcal{T}}}^{\mu\nu}_{\text{gen}}\Big]\,. \tag{8e}$$

For convenience, we defined the observer scalar

$$2\Theta \equiv (p+\hat{\mathcal{V}})^2 - (\hat{\mathcal{S}} - m_\psi)^2 - \hat{\mathcal{A}}^2 - 2\hat{\mathcal{X}} - \hat{\mathcal{P}}^2\,, \tag{8f}$$

which involves each of the five operators. Several remarks are in order. First, this result generalizes the propagator obtained for the spin-degenerate operators $\hat{\mathcal{S}}$, $\hat{\mathcal{V}}$ in Reference [36] and that for the spin-nondegenerate operators $\hat{\mathcal{A}}$, $\hat{\mathcal{T}}$ in Reference [37]. It now applies to the full spectrum of Lorentz-violating operators and is valid also for the nonminimal SME. The propagator reduces to the special results published previously when the corresponding operators are set to zero. Second, all operators of different types are coupled to each other and each contribution of Equation (8) transforms consistently under parity transformations, as expected. For example, each term of $\hat{\mathcal{P}}^{(p)}$ transforms as a pseudoscalar. Third, for vanishing Lorentz violation, we have

$$2\Theta = p^2 - m_\psi^2\,, \quad \hat{\mathcal{S}}^{(p)} = m_\psi(p^2 - m_\psi^2)\,, \quad \hat{\mathcal{P}}^{(p)} = 0\,, \tag{9a}$$

$$\hat{\mathcal{V}}^{(p)\mu} = (p^2 - m_\psi^2)p^\mu\,, \quad \hat{\mathcal{A}}^{(p)\mu} = 0\,, \quad \hat{\mathcal{T}}^{(p)\mu\nu} = 0\,, \quad \Delta = (p^2 - m_\psi^2)^2\,, \tag{9b}$$

whereupon Equation (6) reproduces the standard fermion propagator

$$\mathrm{i}S|_{\text{LV}=0} = \frac{\mathrm{i}(\not{p} + m_\psi\mathbb{1}_4)}{p^2 - m_\psi^2}\,, \tag{10}$$

stated in [35].

4. Modified Neutrino Dispersion Equation

We consider N flavors of modified neutrinos and also include a description of Majorana neutrinos. To do so, the spinor field Ψ_A is constructed as a $2N$-dimensional multiplet of spinors

$$\Psi_A = \begin{pmatrix} \psi_a \\ \psi_a^C \end{pmatrix}\,, \tag{11}$$

where a ranges over N flavors and A labels the $2N$ components of the multiplet. Furthermore, ψ_a^C is the charge conjugate of ψ_a [10]. Due to the form of the construction, there is a redundancy in Ψ encoded in the relationship

$$\Psi^C = \mathcal{C}\Psi\,, \quad \mathcal{C} = \begin{pmatrix} 0 & \mathbb{1}_N \\ \mathbb{1}_N & 0 \end{pmatrix}\,, \tag{12}$$

where the $(2N \times 2N)$ matrix $\mathcal{C}$ is defined in terms of $(N \times N)$ blocks in flavor space. With this information at hand, we present the Lagrange density that incorporates Lorentz and CPT violation into the neutrino sector:

$$\mathcal{L} = \frac{1}{2}\overline{\Psi}_A(\mathrm{i}\not{\partial}\delta_{AB} - M_{AB} + \hat{\mathcal{Q}}_{AB})\Psi_B + \text{H.c.}\,, \tag{13}$$

with the flavor indices A, B. The corresponding Dirac operator $\mathcal{D}_\nu$ in momentum space is a $(8N \times 8N)$ matrix that can be expressed in the form [10]

$$\mathcal{D}_\nu = \mathbb{1}_{2N} \otimes \not{p} - M \otimes \mathbb{1}_4 + \hat{\mathcal{Q}}\,, \tag{14a}$$

where

$$\hat{\mathcal{Q}} = \hat{\mathcal{S}} \otimes \mathbb{1}_4 + \mathrm{i}\hat{\mathcal{P}} \otimes \gamma^5 + \hat{\mathcal{V}}^\mu \otimes \gamma_\mu + \hat{\mathcal{A}}^\mu \otimes \gamma^5\gamma_\mu + \frac{\mathrm{i}}{2}\hat{\mathcal{T}}^{\mu\nu} \otimes \sigma_{\mu\nu} \,. \tag{14b}$$

Here, $\otimes$ denotes a tensor product of $(2N \times 2N)$ matrices in flavor space and matrices in four-dimensional spinor space. Furthermore, M is the $(2N \times 2N)$ neutrino mass matrix. Both the mass matrix and the Lorentz-violating operator $\hat{\mathcal{Q}}$ are expressed in the basis of flavor eigenstates. Therefore, the mass matrix cannot simply be taken as diagonal. The operator $\hat{\mathcal{Q}}$ can be decomposed as in Equation (2), but now we need to remember that the individual contributions also have a flavor structure.

The dispersion equation corresponds to the determinant of the Dirac operator set equal to zero. Therefore, the basic problem is to obtain this determinant and to express the result in a convenient manner. It turns out to be very useful to treat the Dirac operator as a $(2N \times 2N)$ matrix in flavor space where each of these entries on its own is a (4×4) matrix in spinor space. Thus, the structure of this operator is

$$\mathcal{D}_\mathrm{v} = \begin{pmatrix} \mathcal{D}_{1,1} & \dots & \mathcal{D}_{1,2N} \\ \vdots & \ddots & \vdots \\ \mathcal{D}_{2N,1} & \dots & \mathcal{D}_{2N,2N} \end{pmatrix} , \tag{15a}$$

where

$$\mathcal{D}_{i,j} = \delta_{i,j}\not{p} - M_{i,j}\mathbb{1}_4 + \hat{\mathcal{Q}}_{i,j} \,, \tag{15b}$$

$$\hat{\mathcal{Q}}_{i,j} = \left(\hat{\mathcal{S}}\mathbb{1}_4 + \mathrm{i}\hat{\mathcal{P}}\gamma^5 + \hat{\mathcal{V}}^\mu\gamma_\mu + \hat{\mathcal{A}}^\mu\gamma^5\gamma_\mu + \frac{1}{2}\hat{\mathcal{T}}^{\mu\nu}\sigma_{\mu\nu}\right)_{i,j} \,. \tag{15c}$$

The indices stated explicitly are flavor indices, whereby spinor indices are suppressed, as usual. Writing the Dirac operator in this form turns out to be advantageous to apply a sophisticated algorithm for computing the determinant of a matrix in block form in a suggestive way [38]. The algorithm is a recursive method that will be briefly described as follows.

The base is to define $2N$ sets of matrices $\alpha_{i,j}^{(k)}$ with $k \in \{0 \dots 2N-1\}$. Each set for a fixed k contains $(2N)^2$ such matrices, i.e., we have $i, j \in \{1 \dots 2N\}$. To avoid confusion, we emphasize again that all matrices are (4×4), i.e., the indices are not spinor indices, but they label these matrices in flavor space. Now, the first step of the recursion is to assign the Dirac block at the position (i, j) in flavor space to the matrix $\alpha_{i,j}^{(0)}$:

$$\alpha_{i,j}^{(0)} \equiv \mathcal{D}_{i,j} \,. \tag{16}$$

A recurrence relation is defined that allows for constructing new sets of matrices from previous ones. Hence, for $k \in \{0 \dots 2N-2\}$ fixed, we obtain a new set of (4×4) matrices $\{\alpha^{(k+1)}\}$ from the set $\{\alpha^{(k)}\}$ via

$$\alpha_{i,j}^{(k+1)} = \alpha_{i,j}^{(k)} - \alpha_{i,2N-k}^{(k)}(\alpha_{2N-k,2N-k}^{(k)})^{-1}\alpha_{2N-k,j}^{(k)} \,, \tag{17}$$

where A^{-1} denotes the inverse of the matrix A. Having these $2N$ sets of matrices at hand, the determinant of the original block matrix in Equation (15a) is given by

$$\det \mathcal{D}_\mathrm{v} = \prod_{k=1}^{2N} \det(\alpha_{k,k}^{(2N-k)}) \,. \tag{18}$$

Hence, to compute the determinant, only a subset of the matrices obtained before is necessary. This procedure has a great advantage compared to a brute-force evaluation of the determinant. In virtue of Equation (18), the determinant of the full Dirac operator including the flavor structure decomposes into a product of determinants of matrizes that have the form of single-fermion Dirac operators.

Below, we intend to apply this powerful algorithm to the nonminimal neutrino sector. To simplify our notation, we introduce a generic Lorentz-violating operator $\hat{\mathcal{O}}^X$ with a suitable Lorentz index structure X. The latter can stand for one of the five possible operators: $\hat{\mathcal{O}}^X \in \{\hat{\mathcal{S}}, \hat{\mathcal{P}}, \hat{\mathcal{V}}^\mu, \hat{\mathcal{A}}^\mu, \hat{\mathcal{T}}^{\mu\nu}\}$. Now, the steps to be used for the algorithm are as follows:

1. Definition of initial operators:

   According to Equation (16), the first step of the recursion is

$$\alpha_{i,j}^{(0)} = \left(\hat{\mathcal{S}}\mathbb{1}_4 + \mathrm{i}\hat{\mathcal{P}}\gamma^5 + \hat{\mathcal{V}}^\mu\gamma_\mu + \hat{\mathcal{A}}^\mu\gamma^5\gamma_\mu + \frac{1}{2}\hat{\mathcal{T}}^{\mu\nu}\sigma_{\mu\nu}\right)_{i,j}^{(0)}, \tag{19a}$$

   with

$$\hat{\mathcal{S}}_{i,j}^{(0)} = -M_{i,j} + \hat{\mathcal{S}}_{i,j}, \quad \hat{\mathcal{P}}_{i,j}^{(0)} = \hat{\mathcal{P}}_{i,j}, \quad \hat{\mathcal{V}}_{i,j}^{(0)\mu} = p^\mu\delta_{i,j} + \hat{\mathcal{V}}_{i,j}^\mu, \tag{19b}$$

$$\hat{\mathcal{A}}_{i,j}^{(0)\mu} = \hat{\mathcal{A}}_{i,j}^\mu, \quad \hat{\mathcal{T}}_{i,j}^{(0)\mu\nu} = \hat{\mathcal{T}}_{i,j}^{\mu\nu}. \tag{19c}$$

2. Computation of inverse matrix:

   The recurrence relation (17) requires the inverse of $\alpha_{i,j}^{(k)}$, which has the form of a single-fermion Dirac operator. Therefore, its inverse is linked to the fermion propagator obtained in Equation (6) where

$$(\alpha_{2N-k,2N-k}^{(k)})^{-1} = S|_{\hat{\mathcal{O}}^X = \hat{\mathcal{O}}^{(k)X}_{2N-k,2N-k}}. \tag{20a}$$

   The denominator that appears in the latter expression is given by the left-hand side of the dispersion Equation (3):

$$\Delta^{(k)} \equiv \det(\alpha_{2N-k,2N-k}^{(k)}) = \Delta|_{\hat{\mathcal{O}}^X = \hat{\mathcal{O}}^{(k)X}_{2N-k,2N-k}}. \tag{20b}$$

3. Product of the second and third factor of recurrence relation:

   Keeping the previous result in mind, the product necessary to evaluate Equation (17) contains all kinds of combinations of Dirac matrices. As the 16 matrices of the set Γ^A form a basis of (4×4) matrices, these combinations can be completely expressed in terms of the 16 original matrices. To do so, a slew of matrix identities are indispensable, which are to be found in Appendix A. The result of the product then has the following form:

$$(\alpha_{2N-k,2N-k}^{(k)})^{-1}\alpha_{2N-k,j}^{(k)} = \frac{1}{\Delta^{(k)}}\left(\hat{\bar{\mathcal{S}}} + \mathrm{i}\hat{\bar{\mathcal{P}}} + \hat{\bar{\mathcal{V}}}^\mu\gamma_\mu + \hat{\bar{\mathcal{A}}}^\mu\gamma^5\gamma_\mu + \frac{1}{2}\hat{\bar{\mathcal{T}}}^{\mu\nu}\sigma_{\mu\nu}\right)_{2N-k,j}^{(k)}, \tag{21}$$

   according to the single-fermion result of Equation (3). A bar is added to the new operators that follow from this product. These are explicitly given by

$$\hat{\bar{\mathcal{S}}}_{2N-k,j}^{(k)} = \left(\hat{\mathcal{S}}^{(p)}\hat{\mathcal{S}} - \hat{\mathcal{P}}^{(p)}\hat{\mathcal{P}} + \hat{\mathcal{V}}^{(p)}\cdot\hat{\mathcal{V}} - \hat{\mathcal{A}}^{(p)}\cdot\hat{\mathcal{A}} - \frac{1}{2}\hat{\mathcal{T}}^{(p)}\cdot\hat{\mathcal{T}}\right)_{2N-k,j}^{(k)}, \tag{22a}$$

$$\mathrm{i}\hat{\bar{\mathcal{P}}}_{2N-k,j}^{(k)} = \left(\mathrm{i}\hat{\mathcal{S}}^{(p)}\hat{\mathcal{P}} + \mathrm{i}\hat{\mathcal{P}}^{(p)}\hat{\mathcal{S}} - \hat{\mathcal{V}}^{(p)}\cdot\hat{\mathcal{A}} + \hat{\mathcal{A}}^{(p)}\cdot\hat{\mathcal{V}} + \mathrm{i}[\star(\hat{\mathcal{T}}^{(p)}\wedge\hat{\mathcal{T}})]\right)_{2N-k,j}^{(k)}, \tag{22b}$$

$$\begin{aligned}\hat{\bar{\mathcal{V}}}_{2N-k,j}^{(k)\mu} = \Big(&\hat{\mathcal{S}}^{(p)}\hat{\mathcal{V}}^\mu + \hat{\mathcal{V}}^{(p)\mu}\hat{\mathcal{S}} + \mathrm{i}(\hat{\mathcal{V}}^{(p)}\cdot\hat{\mathcal{T}} + \hat{\mathcal{T}}^{(p)}\cdot\hat{\mathcal{V}})^\mu + \mathrm{i}(\hat{\mathcal{P}}^{(p)}\hat{\mathcal{A}}^\mu - \hat{\mathcal{A}}^{(p)\mu}\hat{\mathcal{P}}) \\ &+[\star(\hat{\mathcal{A}}^{(p)}\wedge\hat{\mathcal{T}} + \hat{\mathcal{T}}^{(p)}\wedge\hat{\mathcal{A}})]^\mu\Big)_{2N-k,j}^{(k)},\end{aligned} \tag{22c}$$

$$\hat{\mathcal{A}}^{(k)\mu}_{2N-k,j} = \Big(\hat{\mathcal{S}}^{(p)}\hat{\mathcal{A}}^{\mu} + \hat{\mathcal{A}}^{(p)\mu}\hat{\mathcal{S}} + \mathrm{i}(\hat{\mathcal{P}}^{(p)}\hat{\mathcal{V}}^{\mu} - \hat{\mathcal{V}}^{(p)\mu}\hat{\mathcal{P}}) + \mathrm{i}(\hat{\mathcal{A}}^{(p)}\cdot\hat{\mathcal{T}} + \hat{\mathcal{T}}^{(p)}\cdot\hat{\mathcal{A}})^{\mu}$$
$$+[\star(\hat{\mathcal{V}}^{(p)}\wedge\hat{\mathcal{T}} + \hat{\mathcal{T}}^{(p)}\wedge\hat{\mathcal{V}})]^{\mu}\Big)^{(k)}_{2N-k,j}, \tag{22d}$$

$$\frac{1}{2}\hat{\hat{\mathcal{T}}}^{(k)\mu\nu}_{2N-k,j} = \Big(\frac{1}{2}(\hat{\mathcal{S}}^{(p)}\hat{\mathcal{T}}^{\mu\nu} + \hat{\mathcal{T}}^{(p)\mu\nu}\hat{\mathcal{S}}) - \frac{1}{2}(\hat{\mathcal{P}}^{(p)}\tilde{\hat{\mathcal{T}}}^{\mu\nu} + \tilde{\hat{\mathcal{T}}}^{(p)\mu\nu}\hat{\mathcal{P}}) - \frac{\mathrm{i}}{2}\hat{\mathcal{V}}^{(p)\mu}\wedge\hat{\mathcal{V}}^{\nu}$$
$$+\frac{\mathrm{i}}{2}\hat{\mathcal{A}}^{(p)\mu}\wedge\hat{\mathcal{A}}^{\nu} + \frac{1}{2}[\star(\hat{\mathcal{A}}^{(p)}\wedge\hat{\mathcal{V}} - \hat{\mathcal{V}}^{(p)}\wedge\hat{\mathcal{A}})]^{\mu\nu} + \mathrm{i}(\hat{\mathcal{T}}^{(p)}\cdot\hat{\mathcal{T}})^{[\mu\nu]}\Big)^{(k)}_{2N-k,j}, \tag{22e}$$

where, for brevity, we omit the indices summed over. In the latter results, $\wedge$ stands for the exterior product (wedge product) of two tensors and $\star$ is the Hodge dual of a tensor (cf. Appendix B for a definition of these mathematical operations). We interpret the form of these expressions based on the example of the vector operator of Equation (22c).

First, we will discuss the possible terms that occur. Only certain combinations of basic operators are permitted. In particular, a vector operator can be formed from combinations of the scalar $\hat{\mathcal{S}}$ and the vector $\hat{\mathcal{V}}$. Another possibility is to contract the vector $\hat{\mathcal{V}}$ with the tensor $\hat{\mathcal{T}}$. The fact that a combination of a pseudoscalar $\hat{\mathcal{P}}$ and a pseudovector $\hat{\mathcal{A}}$ transforms as a vector again, explains the third term. Finally, the pseudovector $\hat{\mathcal{A}}$ can be contracted with the tensor $\hat{\mathcal{T}}$ to form a pseudovector. The Hodge dual of the latter provides a vector.

Second, as one of the two operators of each term comes from the inverse (related to the propagator), each of the previously discussed possibilities appears twice. In the second possibility, the roles of the operators simply switch, i.e., what was the vector operator in the first possibility becomes the scalar operator and vice versa.

4. Product of the first factor in recurrence relation and the result of step (3):

   Now we can evaluate the product of the three matrices in Equation (17) completely, which provides

$$\alpha^{(k)}_{i,2N-k}(\alpha^{(k)}_{2N-k,2N-k})^{-1}\alpha^{(k)}_{2N-k,j} = \frac{1}{\Delta^{(k)}}\Big(\hat{\hat{\mathcal{S}}}\mathbb{1}_4 + \mathrm{i}\hat{\hat{\mathcal{P}}}\gamma^5 + \hat{\hat{\mathcal{V}}}^{\mu}\gamma_{\mu} + \hat{\hat{\mathcal{A}}}^{\mu}\gamma^5\gamma_{\mu} + \frac{1}{2}\hat{\hat{\mathcal{T}}}^{\mu\nu}\sigma_{\mu\nu}\Big)^{(k)}_{i,j}. \tag{23}$$

   Following the same procedure as before, we obtain a set of new operators indicated by a double bar. For example, the new vector operator is given by

$$\hat{\hat{\mathcal{V}}}^{(k)\mu}_{i,j} = \Big(\hat{\mathcal{S}}\hat{\hat{\mathcal{V}}}^{\mu} + \hat{\mathcal{V}}^{\mu}\hat{\hat{\mathcal{S}}} + \mathrm{i}(\hat{\mathcal{V}}\cdot\hat{\hat{\mathcal{T}}} + \hat{\mathcal{T}}\cdot\hat{\hat{\mathcal{V}}})^{\mu} + \mathrm{i}(\hat{\mathcal{P}}\hat{\hat{\mathcal{A}}}^{\mu} - \hat{\mathcal{A}}^{\mu}\hat{\hat{\mathcal{P}}})$$
$$+[\star(\hat{\mathcal{A}}\wedge\hat{\hat{\mathcal{T}}} + \hat{\mathcal{T}}\wedge\hat{\hat{\mathcal{A}}})]^{\mu}\Big)^{(k)}_{i,j}. \tag{24}$$

   We see that the structure of the latter result is the same as that of Equation (22c). The simple difference is that the operators have to be renamed according to $\hat{\mathcal{O}}^{(p)X} \mapsto \hat{\mathcal{O}}^{X}$ and $\hat{\mathcal{O}}^{X} \mapsto \hat{\hat{\mathcal{O}}}^{X}$. Analog replacements must be performed for the remaining operators.

5. Recursive step $k \mapsto k+1$:

   Now we have the ingredients to evaluate Equation (17):

$$\alpha^{(k+1)}_{i,j} = \alpha^{(k)}_{i,j} - \alpha^{(k)}_{i,2N-k}(\alpha^{(k)}_{2N-k,2N-k})^{-1}\alpha^{(k)}_{2N-k,j}$$
$$= \Big(\hat{\mathcal{S}}\mathbb{1}_4 + \mathrm{i}\hat{\mathcal{P}}\gamma^5 + \hat{\mathcal{V}}^{\mu}\gamma_{\mu} + \hat{\mathcal{A}}^{\mu}\gamma^5\gamma_{\mu} + \frac{1}{2}\hat{\mathcal{T}}^{\mu\nu}\sigma_{\mu\nu}\Big)^{(k+1)}_{i,j}. \tag{25a}$$

Thus, the $(k+1)$-th operators are expressed in terms of the k-th operators via

$$\mathcal{O}^{(k+1)X}_{i,j} = \mathcal{O}^{(k)X}_{i,j} - \frac{1}{\Delta^{(k)}}\hat{\hat{\mathcal{O}}}^{(k)X}_{i,j} . \tag{25b}$$

6. Express final operators in terms of inicial ones:

   We insert Equation (25b) k times into itself successively to obtain

$$\mathcal{O}^{(k+1)X}_{i,j} = \mathcal{O}^{(0)X}_{i,j} - \sum_{l=0}^{k}\frac{1}{\Delta^{(l)}}\hat{\hat{\mathcal{O}}}^{(l)X}_{i,j} . \tag{26}$$

7. Final computation of determinant:

   All the previous results are employed to compute the determinant according to Equation (18). Doing so, it is reasonable to extract a product of denominators $\Delta^{(n)}$ from the expression such that the determinant itself is a polynomial instead of a sum of fractions of polynomials:

$$\det \mathcal{D}_{\nu} = \prod_{k=1}^{2N}\det(\alpha^{(2N-k)}_{k,k}) = \det(\alpha^{(2N-1)}_{1,1})\prod_{k=2}^{2N}\Delta^{(2N-k)} = \frac{\det(\tilde{\alpha}^{(2N-1)}_{1,1})}{(\Pi^{(2N-2)}_{\Delta})^3} , \tag{27a}$$

   with

$$\Pi^{(k)}_{\Delta} \equiv \prod_{n=0}^{k}\Delta^{(n)} , \tag{27b}$$

$$\tilde{\alpha}^{(2N-1)}_{1,1} \equiv \Pi^{(2N-2)}_{\Delta}\alpha^{(2N-1)}_{1,1} . \tag{27c}$$

   In the forthcoming section, the final result will be stated explicitly.

5. Full Dispersion Equation of Neutrino Sector

As the matrix $\tilde{\alpha}^{(2N-1)}_{1,1}$ contained in the final form of the determinant in Equation (27a) has the same structure as the single-fermion Dirac theory in Equation (1), we can directly compute the full dispersion equation for the nonminimal SME neutrino sector based on the Dirac operator of Equation (14a). The prefactors that we extracted from the determinant in Equation (27a) do not play a role any longer. So neutrinos subject to any kind of Lorentz violation parameterized by the nonminimal SME obey the dispersion equation

$$0 = (\check{\mathcal{S}}^2_- - \check{\mathcal{T}}^2_-)(\check{\mathcal{S}}^2_+ - \check{\mathcal{T}}^2_+) + \check{\mathcal{V}}^2_-\check{\mathcal{V}}^2_+ - 2\check{\mathcal{V}}_- \cdot (\check{\mathcal{S}}_-\eta + 2\mathrm{i}\check{\mathcal{T}}_-)\cdot(\check{\mathcal{S}}_+\eta - 2\mathrm{i}\check{\mathcal{T}}_+)\cdot \check{\mathcal{V}}_+ , \tag{28a}$$

with the operators

$$\check{\mathcal{S}}_{\pm} = (\check{\mathcal{S}} \pm \mathrm{i}\check{\mathcal{P}})^{(2N-1)}_{1,1} , \tag{28b}$$

$$\check{\mathcal{V}}^{\mu}_{\pm} = (\check{\mathcal{V}}^{\mu} \pm \check{\mathcal{A}}^{\mu})^{(2N-1)}_{1,1} , \tag{28c}$$

$$\check{\mathcal{T}}^{\mu\nu}_{\pm} = \frac{1}{2}\left[\check{\mathcal{T}}^{\mu\nu} \pm \mathrm{i}\tilde{\check{\mathcal{T}}}^{\mu\nu}\right]^{(2N-1)}_{1,1} , \tag{28d}$$

and

$$\check{\mathcal{O}}^{(2N-1)X}_{i,j} = \Pi^{(2N-2)}_{\Delta}\mathcal{O}^{(2N-1)X}_{i,j} . \tag{28e}$$

The latter result is the central finding in the current paper. The interesting observation is that the dispersion equation for N neutrino flavors (including the description of both Dirac and Majorana neutrinos) subject to Lorentz violation has a form completely analog to the dispersion equation of the single-fermion sector stated in Equation (3). In general, the coefficients appearing in the dispersion equation are (lengthy) combinations of the neutrino coefficients that are given by a subsequent application of Equations (22) and (24) (for the vector operator, in particular) and Equation (25b). This procedure has to be repeated a sufficient number of times to be able to compute the final necessary operator via Equation (26). The advantage of the result given by Equation (28a) is that it is covariant and does not apply to only a specific observer frame. We think that this form is also suitable to be used in a computer algebra system.

6. First-Order Behavior of Dispersion Relations

As Equation (28a) is, in general, a polynomial in p_0 of high degree, it is challenging to obtain exact dispersion relations from it. Thus, in the current section we intend to get some idea on the general structure of modified neutrino dispersion relations at leading order in Lorentz violation. Since $\hat{\mathcal{S}}^{(k)}_{i,j}$ and $\hat{\mathcal{V}}^{(k)\mu}_{i,j}$ exhibit Lorentz-invariant parts, it is reasonable to decompose the latter operators into two contributions to separate both pieces from each other. By doing so, we get

$$\hat{\mathcal{O}}^{(k)}_{i,j}=\hat{\mathcal{O}}^{(0)}_{i,j}-\sum_{l=0}^{k-1}\frac{1}{\Delta^{(l)}_{2N-l,2N-l}}\hat{\hat{\mathcal{O}}}^{(l)}_{i,j}\approx\hat{\mathcal{O}}^{(k)}_{0;i,j}+\delta\hat{\mathcal{O}}^{(k)}_{i,j}\,. \tag{29a}$$

The additional index "0" (without parentheses) denotes a Lorentz-invariant part. In an analog manner, we generically expand the operators $\hat{\mathcal{P}}^{(k)}_{i,j}$, $\hat{\mathcal{A}}^{(k)\mu}_{i,j}$ and $\hat{\mathcal{T}}^{(k)\mu\nu}_{i,j}$ in the form

$$\hat{\mathcal{O}}^{(k)}_{i,j}=\hat{\mathcal{O}}^{(0)}_{i,j}-\sum_{l=0}^{k-1}\frac{1}{\Delta^{(l)}_{2N-l,2N-l}}\hat{\hat{\mathcal{O}}}^{(l)}_{i,j}\approx\delta\hat{\mathcal{O}}^{(k)}_{i,j}\,. \tag{29b}$$

Here, the notation δ indicates that all leading-order Lorentz-violating contributions are included (recall that $\hat{\mathcal{P}}$, $\hat{\mathcal{A}}$ and $\hat{\mathcal{T}}$ must be expanded to second order). Later on we can substitute those parts by the corresponding expansions. The expanded dispersion equation then takes the form

$$\begin{aligned}0\approx{}&\left[(\hat{\mathcal{S}}^{(k)}_{0;i,j})^2+2\hat{\mathcal{S}}^{(k)}_{0;i,j}\delta\hat{\mathcal{S}}^{(k)}_{i,j}-(\hat{\mathcal{V}}^{(k)}_{0;i,j})^2-2\hat{\mathcal{V}}^{(k)}_{0;i,j}\cdot\delta\hat{\mathcal{V}}^{(k)}_{i,j}\right]^2\\&+2\left[(\hat{\mathcal{S}}^{(k)}_{0;i,j})^2-(\hat{\mathcal{V}}^{(k)}_{0;i,j})^2\right](\delta\hat{\mathcal{P}}^{(k)}_{i,j})^2-4(\delta\hat{\mathcal{Y}}^{(k)}_{i,j})^2\,,\end{aligned} \tag{30a}$$

with the spin-nondegenerate part

$$\begin{aligned}(\delta\hat{\mathcal{Y}}^{(k)}_{i,j})^2={}&(\hat{\mathcal{V}}^{(k)}_{0;i,j}\cdot\delta\hat{\mathcal{A}}^{(k)}_{i,j})^2-\frac{1}{2}\left[(\hat{\mathcal{V}}^{(k)}_{0;i,j})^2+(\hat{\mathcal{S}}^{(k)}_{0;i,j})^2\right](\delta\hat{\mathcal{A}}^{(k)}_{i,j})^2+2\hat{\mathcal{S}}^{(k)}_{0;i,j}\hat{\mathcal{V}}^{(k)}_{0;i,j}\cdot\delta\hat{\tilde{\mathcal{T}}}^{(k)}_{i,j}\cdot\delta\hat{\mathcal{A}}^{(k)}_{i,j}\\&+\hat{\mathcal{V}}^{(k)}_{0;i,j}\cdot\delta\hat{\tilde{\mathcal{T}}}^{(k)}_{i,j}\cdot\delta\hat{\tilde{\mathcal{T}}}^{(k)}_{i,j}\cdot\hat{\mathcal{V}}^{(k)}_{0;i,j}+\frac{1}{4}\left[(\hat{\mathcal{V}}^{(k)}_{0;i,j})^2-(\hat{\mathcal{S}}^{(k)}_{0;i,j})^2\right](\delta\hat{\tilde{\mathcal{T}}}^{(k)}_{i,j})^2\,.\end{aligned} \tag{30b}$$

Note that the last term on the right-hand side of the latter equation vanishes for the single-fermion sector. The leading-order expansion of the dispersion equation can be further expressed as

$$\begin{aligned}(\hat{\mathcal{V}}^{(k)}_{0;i,j})^2-(\hat{\mathcal{S}}^{(k)}_{0;i,j})^2\approx{}&2\hat{\mathcal{S}}^{(k)}_{0;i,j}\delta\hat{\mathcal{S}}^{(k)}_{i,j}-2\hat{\mathcal{V}}^{(k)}_{0;i,j}\cdot\delta\hat{\mathcal{V}}^{(k)}_{i,j}\\&\pm2\sqrt{(\delta\hat{\mathcal{Y}}^{(k)}_{i,j})^2-\frac{1}{2}\left[(\hat{\mathcal{S}}^{(k)}_{0;i,j})^2-(\hat{\mathcal{V}}^{(k)}_{0;i,j})^2\right](\delta\hat{\mathcal{P}}^{(k)}_{i,j})^2}\,.\end{aligned} \tag{31}$$

It shall be emphasized again that the Lorentz-invariant pieces are contained only in $\hat{\mathcal{V}}^{(k)}_{0;i,j}$ and $\hat{\mathcal{S}}^{(k)}_{0;i,j}$, respectively. The term on the left-hand side of the equation above is a polynomial of order $8N$ in p^0

and contains the pure-mass part. The standard dispersion relations for the case of an arbitrary number of N flavors seem to follow the pattern

$$E_0^{(u)} \stackrel{?}{=} \sqrt{\mathbf{p}^2 + \frac{1}{2N}\left[\mathfrak{M} + (m_{\text{eff}}^{(u)})^2\right]}\,, \quad u \in \{1\dots 2N\}\,, \tag{32a}$$

where $\mathbf{p}$ is the spatial momentum of p^μ and

$$\mathfrak{M} = \text{Tr}(M^2) = \sum_{i,j=1}^{2N} M_{i,j}M_{j,i}\,. \tag{32b}$$

Furthermore, $(m_{\text{eff}}^{(u)})^2$ for u fixed is a complicated polynomial of mass matrix coefficients that we simply called an "effective mass squared." The validity of Equation (32a) is challenging to proof for an arbitrary N, though. Taking Lorentz violation into account, the first-order modification is given by

$$\begin{aligned} E^{(u)} \approx E_0^{(u)} + \frac{1}{2NE_0^{(u)}(m_{\text{eff}}^{(u)})^2}&\Bigg\{\hat{\mathcal{S}}_{0;i,j}^{(k)}\delta\hat{\mathcal{S}}_{i,j}^{(k)} - \hat{\mathcal{V}}_{0;i,j}^{(k)}\cdot\delta\hat{\mathcal{V}}_{i,j}^{(k)} \\ &\pm\sqrt{(\delta\hat{\mathcal{Y}}_{i,j}^{(k)})^2 - \frac{1}{2}\left[(\hat{\mathcal{S}}_{0;i,j}^{(k)})^2 - (\hat{\mathcal{V}}_{0;i,j}^{(k)})^2\right](\delta\hat{\mathcal{P}}_{i,j}^{(k)})^2}\Bigg\}\,, \end{aligned} \tag{33}$$

with the operators defined in Equations (29a), (29b) and (30b). Here we see how the presence of the spin-nondegenerate operators doubles the number of dispersion relations, as expected. In total, there are then $8N$ modified dispersion laws, which corresponds to the degree of the polynomial in p^0. Furthermore, the pseudoscalar operator also leads to such a doubling. If the pseudo-scalar operator is the only source for Lorentz violation, it contributes at second order, as $\delta\hat{\mathcal{P}}_{i,j}^{(k)}$ is of second order in Lorentz violation.

Special Case: $N = 1$

As even the general first-order expansion is quite complicated, it shall be exemplified as follows. We consider the theory of a single neutrino flavor that can be of either Dirac or Majorana type. It is reasonable to switch Lorentz violation off at first. The dispersion equation is then a polynomial of fourth degree. (In principle, the polynomial on the right-hand side of the dispersion equation is raised to the second power, that is, the degeneracy of all zeros is doubled.) It reads:

$$0 = p^4 - \mathfrak{M}p^2 + \tilde{\mathfrak{M}}\,, \tag{34a}$$

where

$$\mathfrak{M} = \text{Tr}(M^2)\,, \quad \tilde{\mathfrak{M}} = (\det M)^2\,. \tag{34b}$$

Solving for p_0 delivers two distinct energies:

$$E_0^{(1,2)} = \sqrt{\mathbf{p}^2 + \frac{\mathfrak{M}}{2} \pm \sqrt{\left(\frac{\mathfrak{M}}{2}\right)^2 - \tilde{\mathfrak{M}}}}\,. \tag{35}$$

Now, the neutrino energies modified by Lorentz violation are found to have the form

$$E^{(1,2)\pm} = E_0^{(1,2)} - \frac{1}{2E_0^{(1,2)}}\left(\hat{\mathcal{S}}_{\text{disp}}^{(1,2)} + p\cdot\hat{\mathcal{V}}_{\text{disp}}^{(1,2)} \pm \delta\hat{\mathcal{Y}}\right) + \dots\,, \tag{36a}$$

with

$$\hat{\mathcal{S}}^{(1,2)}_{\mathrm{disp}} = M_{ab}\hat{\mathcal{S}}_{ba} \pm \frac{(M_{aa}+M_{bb})M_{ab}\hat{\mathcal{S}}_{ba} - (M_{aa}^2 + M_{aa}M_{bb} - 2M_{ab}M_{ba})\hat{\mathcal{S}}_{aa}}{\sqrt{(M_{11}-M_{22})^2 + 4M_{12}M_{21}}}\,, \tag{36b}$$

$$\hat{\mathcal{V}}^{(1,2)\mu}_{\mathrm{disp}} = \hat{\mathcal{V}}^{\mu}_{aa} \pm \frac{2M_{ab}\hat{\mathcal{V}}^{\mu}_{ba} - M_{aa}\hat{\mathcal{V}}^{\mu}_{bb}}{\sqrt{(M_{11}-M_{22})^2 + 4M_{12}M_{21}}}\,, \tag{36c}$$

$$\begin{aligned}(\delta\hat{\mathcal{Y}})^2 &= \frac{1}{(M_{11}+M_{22})^2\left[(M_{11}-M_{22})^2+4M_{12}M_{21}\right]} \\ &\quad\times\left\{(\hat{\mathcal{V}}_0\cdot\delta\hat{\mathcal{A}})^2 - \frac{1}{2}\left[(\hat{\mathcal{V}}_0)^2+(\hat{\mathcal{S}}_0)^2\right](\delta\hat{\mathcal{A}})^2 + 2\hat{\mathcal{S}}_0\hat{\mathcal{V}}_0\cdot\delta\hat{\tilde{\mathcal{T}}}\cdot\delta\hat{\mathcal{A}}\right. \\ &\quad\left.+\hat{\mathcal{V}}_0\cdot\delta\hat{\tilde{\mathcal{T}}}\cdot\delta\hat{\tilde{\mathcal{T}}}\cdot\hat{\mathcal{V}}_0 + \frac{1}{4}\left[(\hat{\mathcal{V}}_0)^2-(\hat{\mathcal{S}}_0)^2\right](\delta\hat{\tilde{\mathcal{T}}})^2\right\}^{(1)}_{1,1}.\end{aligned} \tag{36d}$$

Furthermore,

$$\delta\hat{\mathcal{A}}^{(1)\mu}_{1,1} \approx \hat{\mathcal{A}}^{(1)\mu}_{1,1} = \hat{\mathcal{A}}^{(0)\mu}_{1,1} - \frac{1}{\Delta^{(0)}_{2,2}}\hat{\hat{\mathcal{A}}}^{(0)\mu}_{1,1}\,, \tag{37a}$$

$$\delta\hat{\mathcal{T}}^{(k)\mu\nu}_{1,1} \approx \hat{\mathcal{T}}^{(1)\mu\nu}_{1,1} = \hat{\mathcal{T}}^{(0)\mu\nu}_{1,1} - \frac{1}{\Delta^{(0)}_{2,2}}\hat{\hat{\mathcal{T}}}^{(0)\mu\nu}_{1,1}\,, \tag{37b}$$

and

$$\hat{\mathcal{S}}_0 = -M_{11} - \frac{M_{12}M_{22}M_{21}}{p^2 - M_{22}}\,, \tag{37c}$$

$$\hat{\mathcal{V}}^{\mu}_0 = \frac{p^2 - M_{12}M_{21} - M_{11}^2}{p^2 - M_{22}}p^{\mu}\,. \tag{37d}$$

The flavor indices in Equations (36b) and (36c) are understood to be summed over. Note that in Equation (36a) two signs can be chosen at different positions independently from each other. The first sign is indicated by the suffices (1,2) and appears in the Lorentz-invariant and spin-degenerate parts of the dispersion relation. The second sign is marked by the additional index $\pm$ and is related to the spin-nondegenerate coefficients only, which are incorporated in the quantity $(\delta\hat{\mathcal{Y}})^2$. Hence, for $N = 1$ there can be already 4 different dispersion relations. With spin-nondegenerate Lorentz violation present, there are two distinct dispersion relations for each of the two neutrino types. The number of modified dispersion laws is supposed to increase with the number of neutrino flavors.

7. Classical Lagrangians

As a final application of all the previous results, we want to map the neutrino field theory to a theory of classical, relativistic, pointlike particles. The latter is described by a Lagrange function in terms of the four-velocity u^{μ}. The technique to carry out such a mapping in the context of the SME was developed in Reference [39] and has been subject to intense studies for the past ten years. Presently, the procedure is well-known and has been applied to both the minimal and nonminimal SME fermion sector. At leading order in Lorentz violation, it was demonstrated that a classical Lagrangian can be obtained from the dispersion relation directly via a mapping procedure [40]. We will employ this method for the case $N = 1$, which leads to the classical Lagrangians corresponding to this field theory of modified neutrinos. The Lagrangians can be cast into the form

$$L^{(1,2)\pm} = -M^{(1,2)}\sqrt{u^2}\left[1 - \frac{(\hat{\mathcal{S}}^{(1,2)}_{\mathrm{disp}})_* \pm \delta\hat{\mathfrak{Y}}_*}{2(M^{(1,2)})^2} - \frac{u\cdot(\hat{\mathcal{V}}^{(1,2)}_{\mathrm{disp}})_*}{2M^{(1,2)}\sqrt{u^2}} + \dots\right]\,, \tag{38a}$$

$$M^{(1,2)} = \sqrt{\frac{\mathfrak{M}}{2} \pm \sqrt{\left(\frac{\mathfrak{M}}{2}\right)^2 - \tilde{\mathfrak{M}}}\,, \tag{38b}$$

with $\mathfrak{M}$ and $\tilde{\mathfrak{M}}$ of Equation (34b). The quantities endowed with an asterisk emerge from the expressions defined in Equation (36) in replacing each Lorentz-violating operator by a suitable contraction of the corresponding controlling coefficients with four-velocities. Each contraction involves an additional prefactor that depends on the mass dimension. For a generic Lorentz-violating operator of mass dimension d, this contraction reads

$$(\hat{\mathcal{O}}^{(k)}_{i,j})_* = \left(\frac{M^{(1,2)}}{\sqrt{u^2}}\right)^{d-3}\mathcal{O}^{(k)\alpha_1\dots\alpha_{d-3}}_{i,j}u_{\alpha_1}\dots u_{\alpha_{d-3}}\,. \tag{39}$$

As there are four modified dispersion relations, there are also four classical Lagrangians. Each of those describes a type of massive "classical neutrino." For vanishing Lorentz violation, the Lagrangians take the form $L^{(1,2)} = -M^{(1,2)}\sqrt{u^2}$, which is analogous to the standard result $L = -m_\psi\sqrt{u^2}$ for a single Dirac fermion of mass m_ψ under the identification $m_\psi = M^{(1,2)}$. Here we see how the combinations $M^{(1,2)}$ of mass matrix coefficients can be interpreted as something like a simple mass of the classical neutrino analog. This behavior is a classical remnant of the quantum effect of neutrino mixing. Furthermore, the classical Lagrangian can be checked to be positively homogeneous of degree 1 in u^μ, as expected.

Classical Lagrangians as those of Equation (38) are valuable in the description of Lorentz violation for neutrinos in the presence of an external gravitational field. Since the current section serves only as a demonstration of the basic procedure, we will not delve deeper into this interesting topic. Whenever the neutrino masses are simply neglected, such classical Lagrangians simply lose their meaning. Neutrino propagation in a gravitational background must then be described with a different formalism such as the eikonal equation, which turned out to be fruitful for massless particles, e.g., photons [41].

8. Conclusions

In this paper we derived the full propagator of a single-fermion Dirac theory based on the nonminimal SME as well as the full dispersion equation of modified neutrinos. Both results are expressed in a covariant form. Although it is quite an essential tool in perturbation theory, the full propagator of the nonminimal SME fermion sector has not been stated elsewhere, so far. The dispersion equation is valid for the general case of N neutrino flavors with the description of both Dirac and Majorana neutrinos included. Despite the additional flavor structure and the distinction between Dirac and Majorana neutrinos, we found that the dispersion equation has a structure analogous to that of a single Dirac fermion. However, it is also clear that the form of the Lorentz-violating operators that occur in the neutrino dispersion equation is much more involved because of the additional flavor structure. We also investigated the dispersion relations at leading order in Lorentz violation for basic configurations. Finally, we included a brief comment on classical Lagrangians in the context of neutrinos. Our findings are technical, but they may be valuable in forthcoming phenomenological works on Lorentz-violating Dirac fermions and neutrinos.

Author Contributions: Formal analysis, J.A.A.d.S.d.R. and M.S.; writing—original draft preparation, M.S.; writing—review and editing, J.A.A.d.S.d.R. and M.S.

Funding: This research was funded by CNPq grant numbers Universal 421566/2016-7, Produtividade 312201/2018-4 and by FAPEMA grant number Universal 01149/17.

Acknowledgments: The authors are grateful for financial support by the Brazilian agencies CNPq, CAPES, and FAPEMA. We thank the two anonymous referees for a number of constructive comments on the first submitted version of the manuscript.

Conflicts of Interest: The authors declare no conflict of interest.

Appendix A. Useful Relations for Dirac Matrices

For future reference, we think that it is a good idea to list the relations for Dirac matrices that we used to obtain our results. First, it is difficult to find the whole set of relations at a particular place. Second, in some works we even encountered typos. Therefore, the validity of the following relations was checked explicitly and they are supposed to be correct, as they stand:

$$\gamma_\mu\gamma_\nu = \eta_{\mu\nu}\mathbb{1}_4 - \mathrm{i}\sigma_{\mu\nu}\,, \tag{A1a}$$

$$\sigma_{\mu\nu}\gamma^5 = \frac{\mathrm{i}}{2}\varepsilon_{\alpha\beta\mu\nu}\sigma^{\alpha\beta}\,, \tag{A1b}$$

$$\gamma_\mu\gamma_\nu\gamma^5 = \eta_{\mu\nu}\gamma^5 + \frac{1}{2}\varepsilon_{\mu\nu\alpha\beta}\sigma^{\alpha\beta}\,, \tag{A1c}$$

$$\gamma_\mu\gamma_\nu\gamma_\lambda = \eta_{\mu\nu}\gamma_\lambda + \eta_{\nu\lambda}\gamma_\mu - \eta_{\lambda\mu}\gamma_\nu + \mathrm{i}\varepsilon_{\mu\nu\lambda\rho}\gamma^\rho\gamma^5\,, \tag{A1d}$$

$$\sigma_{\mu\nu}\gamma_\lambda = \mathrm{i}(\eta_{\nu\lambda}\gamma_\mu - \eta_{\lambda\mu}\gamma_\nu) - \varepsilon_{\mu\nu\lambda\rho}\gamma^\rho\gamma^5\,, \tag{A1e}$$

$$\gamma_\mu\sigma_{\nu\lambda} = \mathrm{i}(\eta_{\mu\nu}\gamma_\lambda - \eta_{\lambda\mu}\gamma_\nu) - \varepsilon_{\mu\nu\lambda\rho}\gamma^\rho\gamma^5\,, \tag{A1f}$$

$$\gamma^5\sigma_{\mu\nu}\gamma_\lambda = \mathrm{i}(\eta_{\nu\lambda}\gamma^5\gamma_\mu - \eta_{\lambda\mu}\gamma^5\gamma_\nu) + \varepsilon_{\mu\nu\lambda\rho}\gamma^\rho\,, \tag{A1g}$$

$$\gamma^5\gamma_\mu\sigma_{\nu\lambda} = \mathrm{i}(\eta_{\mu\nu}\gamma^5\gamma_\lambda - \eta_{\lambda\mu}\gamma^5\gamma_\nu) + \varepsilon_{\mu\nu\lambda\rho}\gamma^\rho\,, \tag{A1h}$$

$$[\sigma_{\mu\nu},\sigma_{\lambda\alpha}] = 2\mathrm{i}(\eta_{\mu\lambda}\sigma_{\alpha\nu} - \eta_{\nu\lambda}\sigma_{\alpha\mu} + \eta_{\nu\alpha}\sigma_{\lambda\mu} - \eta_{\mu\alpha}\sigma_{\lambda\nu})\,, \tag{A1i}$$

$$\{\sigma_{\mu\nu},\sigma_{\lambda\alpha}\} = 2\left[(\eta_{\alpha\nu}\eta_{\lambda\mu} - \eta_{\alpha\mu}\eta_{\lambda\nu})\mathbb{1}_4 + \mathrm{i}\varepsilon_{\mu\nu\lambda\alpha}\gamma^5\right]\,, \tag{A1j}$$

$$\sigma_{\mu\nu}\sigma_{\lambda\alpha} = \mathrm{i}(\eta_{\mu\lambda}\sigma_{\alpha\nu} - \eta_{\nu\lambda}\sigma_{\alpha\mu} + \eta_{\nu\alpha}\sigma_{\lambda\mu} - \eta_{\mu\alpha}\sigma_{\lambda\nu}) + (\eta_{\alpha\nu}\eta_{\lambda\mu} - \eta_{\alpha\mu}\eta_{\lambda\nu})\mathbb{1}_4 + \mathrm{i}\varepsilon_{\mu\nu\lambda\alpha}\gamma^5\,. \tag{A1k}$$

We employed these results mainly to obtain Equations (21) and (23).

Appendix B. Definition of Wedge Product and Hodge Dual

The wedge product and Hodge dual are concepts that are of wide use in algebra. They turned out to be fruitful to express Equation (22) in a relatively compact form. We define the wedge product of two (contravariant) tensors A and B as the antisymmetrized direct product of these tensors:

$$\frac{N_a!N_b!}{(N_a+N_b)!}A^{\mu\nu\dots}\wedge B^{\alpha\beta\dots} \equiv A^{[\mu\nu\dots}B^{\alpha\beta\dots]} \equiv C^{\mu\nu\dots\alpha\beta\dots}\,, \tag{A2}$$

where $N_{a,b}$ is the number of indices of the tensor A and B, respectively. The latter product gives rise to a new tensor C with the union of Lorentz indices of the tensors A and B. The definition of the Hodge dual of a tensor depends on the number of dimensions of the space considered. As we work in four-dimensional Minkowski spacetime, the Hodge dual of a covariant two-tensor A provides a contravariant two tensor:

$$(\star A)^{\alpha\beta} \equiv \frac{1}{2} A_{\mu\nu} \varepsilon^{\mu\nu\alpha\beta} . \tag{A3}$$

On the other hand, the Hodge dual of a covariant three-tensor B gives a contravariant vector and that of a four-tensor C is a scalar:

$$(\star B)^{\varrho} \equiv \frac{1}{3!} B_{\mu\nu\lambda} \varepsilon^{\mu\nu\lambda\varrho} , \quad \star C \equiv \frac{1}{4!} C_{\mu\nu\lambda\alpha} \varepsilon^{\mu\nu\lambda\alpha} . \tag{A4}$$

Note that we follow the convention of the prefactors used in mathematics.

References

1. Hagiwara, K.; Hikasa, K.; Nakamura, K.; Tanabashi, M.; Aguilar-Benitez, M.; Amsler, C.; Barnett, R.; Burchat, P.R.; Carone, C.D.; Particle Data Group; et al. Review of particle physics. *Phys. Rev. D* **2018**, *98*, 030001.
2. Bilenky, S.M. The history of neutrino oscillations. *Phys. Scripta T* **2005**, *121*, 17. [CrossRef]
3. Kostelecký, V.A.; Samuel, S. Spontaneous breaking of Lorentz symmetry in string theory. *Phys. Rev. D* **1989**, *39*, 683. [CrossRef] [PubMed]
4. Kostelecký, V.A.; Potting, R. Expectation values, Lorentz invariance, and *CPT* in the open bosonic string. *Phys. Lett. B* **1996**, *381*, 89. [CrossRef]
5. Kostelecký, V.A.; Russell, N. Data tables for Lorentz and *CPT* violation. *Rev. Mod. Phys.* **2011**, *83*, 11. [CrossRef]
6. Greenberg, O.W. *CPT* violation implies violation of Lorentz invariance. *Phys. Rev. Lett.* **2002**, *89*, 231602. [CrossRef]
7. Colladay, D.; Kostelecký, V.A. *CPT* violation and the standard model. *Phys. Rev. D* **1997** *55*, 6760. [CrossRef]
8. Colladay, D.; Kostelecký, V.A. Lorentz-violating extension of the standard model. *Phys. Rev. D* **1998**, *58*, 116002. [CrossRef]
9. Kostelecký, V.A.; Mewes, M. Electrodynamics with Lorentz-violating operators of arbitrary dimension. *Phys. Rev. D* **2009**, *80*, 015020. [CrossRef]
10. Kostelecký, V.A.; Mewes, M. Neutrinos with Lorentz-violating operators of arbitrary dimension. *Phys. Rev. D* **2012**, *85*, 096005. [CrossRef]
11. Kostelecký, V.A.; Mewes, M. Fermions with Lorentz-violating operators of arbitrary dimension. *Phys. Rev. D* **2013**, *88*, 096006. [CrossRef]
12. Katori, T.; Kostelecký, V.A.; Tayloe, R. Global three-parameter model for neutrino oscillations using Lorentz violation. *Phys. Rev. D* **2006**, *74*, 105009. [CrossRef]
13. Díaz, J.S.; Kostelecký, V.A.; Mewes, M. Perturbative Lorentz and *CPT* violation for neutrino and antineutrino oscillations. *Phys. Rev. D* **2009**, *80*, 076007.
14. Katori, T.; MiniBooNE Collaboration. Test for Lorentz and *CPT* Violation with the MiniBooNE Low-Energy Excess. In *CPT and Lorentz Symmetry*, 5th ed.; Kostelecký, V.A., Ed.; World Scientific: Singapore, 2010.
15. Díaz, J.S.; Kostelecký, V.A. Three-parameter Lorentz-violating texture for neutrino mixing. *Phys. Lett. B* **2011**, *700*, 25. [CrossRef]
16. Díaz, J.S.; Kostelecký, V.A. Lorentz- and *CPT*-violating models for neutrino oscillations. *Phys. Rev. D* **2012**, *85*, 016013. [CrossRef]
17. Aguilar-Arevalo, A.A.; Anderson, C.E.; Bazarko, A.O.; Brice, S.J.; Brown, B.C.; Bugel, L.; Cao, J.; Coney, L.; Conrad, J.M.; MiniBooNE Collaboration; et al. Test of Lorentz and *CPT* violation with short baseline neutrino oscillation excesses. *Phys. Lett. B* **2013**, *718*, 1303. [CrossRef]
18. Katori, T.; Kostelecký, V.A.; Tayloe, R. Global three-parameter model for neutrino oscillations using Lorentz violation. *Nucl. Phys. Proc. Suppl.* **2011**, *221*, 357. [CrossRef]
19. Abe, Y.; Aberle, C.; Dos Anjos, J.C.; Bergevin, M.; Bernstein, A.; Bezerra, T.J.; Bezrukhov, L.; Blucher, E.; Bowden, N.S.; Double Chooz Collaboration; et al. First test of Lorentz violation with a reactor-based antineutrino experiment. *Phys. Rev. D* **2012**, *86*, 112009. [CrossRef]

20. Katori, T. Tests of Lorentz and *CPT* violation with MiniBooNE neutrino oscillation excesses. *Mod. Phys. Lett. A* **2012**, *27*, 1230024. [CrossRef]
21. Díaz, J.S.; Kostelecký, V.A.; Lehnert, R. Relativity violations and beta decay. *Phys. Rev. D* **2013**, *88*, 071902.
22. Díaz, J.S.; Kostelecký, V.A.; Mewes, M. Testing relativity with high-energy astrophysical neutrinos. *Phys. Rev. D* **2014**, *89*, 043005.
23. Díaz, J.S. Limits on Lorentz and *CPT* violation from double beta decay. *Phys. Rev. D* **2014**, *89*, 036002. [CrossRef]
24. Díaz, J.S. Neutrinos as probes of Lorentz invariance. *Adv. High Energy Phys.* **2014**, *2014*, 962410. [CrossRef]
25. Díaz, J.S. Tests of Lorentz symmetry in single beta decay. *Adv. High Energy Phys.* **2014**, *2014*, 305298. [CrossRef]
26. Díaz, J.S. Correspondence between nonstandard interactions and *CPT* violation in neutrino oscillations. *arXiv* **2015**, arXiv:1506.01936.
27. Díaz, J.S.; Klinkhamer, F.R. Neutrino refraction by the cosmic neutrino background. *Phys. Rev. D* **2016**, *93*, 053004. [CrossRef]
28. Díaz, J.S.; Schwetz, T. Limits on *CPT* violation from solar neutrinos. *Phys. Rev. D* **2016**, *93*, 093004. [CrossRef]
29. Díaz, J.S. Testing Lorentz and *CPT* invariance with neutrinos. *Symmetry* **2016**, *8*, 105. [CrossRef]
30. Argüelles, C.A.; Collin, G.H.; Conrad, J.M.; Katori, T.; Kheirandish, A. Search for Lorentz violation in km^3-scale neutrino telescopes. In *CPT and Lorentz Symmetry*, 7th ed.; Kostelecký, V.A., Ed.; World Scientific: Singapore, 2017.
31. Katori, T.; Argüelles, C.A.; Salvado, J. Test of Lorentz violation with astrophysical neutrino flavor. In *CPT and Lorentz Symmetry*, 7th ed.; Kostelecký, V.A., Ed.; World Scientific: Singapore, 2017.
32. Abe, K.; Amey, J.; Andreopoulos, C.; Antonova, M.; Aoki, S.; Ariga, A.; Assylbekov, S.; Autiero, D.; Ban, S.; T2K Collaboration; et al. Search for Lorentz and *CPT* violation using sidereal time dependence of neutrino flavor transitions over a short baseline. *Phys. Rev. D* **2017**, *95*, 111101. [CrossRef]
33. Aartsen, M.G.; IceCube Collaboration. Neutrino interferometry for high-precision tests of Lorentz symmetry with IceCube. *Nature Phys.* **2018**, *14*, 961.
34. Kostelecký, V.A.; Lehnert, R. Stability, causality, and Lorentz and *CPT* violation. *Phys. Rev. D* **2001**, *63*, 065008. [CrossRef]
35. Peskin, M.E.; Schroeder, D.V. *An Introduction to Quantum Field Theory*; Perseus Books Publishing LLC: Reading, MA, USA, 1995.
36. Schreck, M. Vacuum Cherenkov radiation for Lorentz-violating fermions. *Phys. Rev. D* **2017**, *96*, 095026. [CrossRef]
37. Reis, J.A.A.S.; Schreck, M. Lorentz-violating modification of Dirac theory based on spin-nondegenerate operators. *Phys. Rev. D* **2017**, *95*, 075016. [CrossRef]
38. Powell, P.D. Calculating determinants of block matrices. *arXiv* **2011**, arXiv:1112.4379.
39. Kostelecký, V.A.; Russell, N. Classical kinematics for Lorentz violation. *Phys. Lett. B* **2010**, *693*, 443. [CrossRef]
40. Reis, J.A.A.S.; Schreck, M. Leading-order classical Lagrangians for the nonminimal standard-model extension. *Phys. Rev. D* **2018**, *97*, 065019. [CrossRef]
41. Schreck, M. Eikonal approximation, Finsler structures, and implications for Lorentz-violating photons in weak gravitational fields. *Phys. Rev. D* **2015**, *92*, 125032. [CrossRef]

symmetry

MDPI

Article

Antimatter Quantum Interferometry

Marco Giammarchi

Istituto Nazionale di Fisica Nucleare—Sezione di Milano, 20133 Milano, Italy; marco.giammarchi@mi.infn.it

Received: 3 September 2019; Accepted: 29 September 2019; Published: 5 October 2019

Abstract: The wave–particle duality hypothesis for massive particles has been confirmed by an overwhelming variety of indirect experimental evidence. In addition, direct interferometric tests have been made on particles like electrons, neutrons and even a few molecules, explicitly showing wave-like diffraction and interference phenomena. Of particular interest in this direction, single particle interference has also been demonstrated, but only for the electron case. No such kind of direct information was available for antiparticles or antimatter in general. After briefly discussing the subjects of antimatter research and interferometry, I present here the first evidence of single particle antimatter interference, made with positrons.

Keywords: quantum mechanics; antimatter; interferometry

---

1. Introduction

The wave–particle duality hypothesis for massive particles was introduced by de Broglie almost a century ago: The Planck constant h relates the momentum p of a massive particle to its de Broglie wavelength: $\lambda_{dB} = h/p$ [1]. This relation, together with the uncertainty principle and the superposition principle, is at the heart of quantum mechanics. These principles have now been tested in an overwhelming variety of experiments over more than 100 years.

Of particular interest is the direct evidence of wave-like behavior of quantum massive particles showing diffraction and interference phenomena, for the first time with electrons [2,3]. Neutrons were shown to display wave behavior in crystals [4], in the gravitational Colella–Overhauser–Werner set of experiments [5,6] and later on using single and double slit diffraction [7]. Wave-like behavior is nowadays established also for molecules like Na_2 [8], and up to the complexity of fullerene [9].

Among the direct tests of wave-like nature of massive particles, a special place is held by experiments where a single particle propagates through an interferometric system. According to Feynman, this ideal experiment constitutes a decisive proof, a test "that has in it the heart of quantum mechanics" [10].

Single-particle experiments were conducted for the first time with electrons in 1976 by G. Merli, G.F. Missiroli and G. Pozzi, in a configuration featuring an electronic biprism, equivalent to the double slit suggested by Feynman [11]. Several decades later, the same experiment was also realized with material slits [12]. At the same time, no direct information on antiparticle wave properties was detected, with the only exception that of an indication of positron diffraction [13].

2. Antimatter

Antimatter, introduced following the Dirac equation in 1928 [14], was observed a few years later in cosmic rays [15]. The general relation between particle and antiparticle properties is the CPT (charge-parity-time reversal) theorem [16], that holds for quantum field theories in a flat spacetime.

While antiparticles are routinely produced by cosmic (and man-made) accelerators, their presence in our environment is negligible and experimentation always requires dedicated sources. CPT symmetry can be studied in principle on any existing antiparticle; however, neutral antimatter (or symmetric matter–antimatter) systems are of particular interest because of the possibility of testing the weak equivalence principle (WEP). The production of cold anti-hydrogen atoms at the CERN Antiproton Decelerator [17,18] has been a milestone in this direction, followed by anti-hydrogen confinement [19] and the development of an antiatom beam [20].

The simplest and most symmetric matter–antimatter system, positronium (Ps, the e^+ e^- bound system), was discovered by M. Deutsch in 1951 [21]. It is constituted by an electron and a positron and has been the subject of intense investigation in the last decades, holding the promise to allow tests of fundamental laws (see [22] and references therein).

In addition to searching for violations of fundamental laws per se, antimatter studies are relevant to the goal of understanding the fundamental baryonic and leptonic asymmetry in the Universe [23]. The most natural mechanism that could predict the asymmetry relies in fact on the Sakharov conditions [24] being verified at the grand unification scale of energy ($\approx$ 10^{16} GeV) and their possible interplay with CPT conservation [25]. Antimatter studies (at both low and high energies) might be necessary to solve this fundamental riddle, related to our own very existence.

3. Antimatter Interferometry

In spite of all the progress in studies about antimatter, no experiments featuring antiparticle interference have ever been done. Preliminary ideas about interferometry for antimatter were considered mainly in the frame of inertial sensing and possible measurements of gravitation for antimatter [26].

Generally speaking, antimatter poses a special problem because of its paucity in terrestrial and astronomical environments. For instance, the antiproton-to-proton ratio in cosmic rays is about 10^{-5} and virtually no antiparticle can survive in the environment because of immediate annihilation with ordinary matter. For these reasons, controlled sources of antiparticles are restricted to high-energy accelerators or radioactive sources. Interferometry also requires antimatter at relatively low energies, suitable for controlled propagation or even confinement, as is the case for the above-mentioned Antiproton Decelerator or the radioactive ^{22}Na positron sources.

Considering the case of positrons, for instance, the available radioactive sources and the following treatment necessary to lower the energy (generally in the *keV* range) results in beam intensities of the order of 10^4 particles per second. In comparison, electron sources can easily reach the mA range—11 orders of magnitude higher!

The QUPLAS (Quantum interferometry and gravitation with positronium) research group has undertaken a systematic program of study on positrons and positronium, whose first step, called QUPLAS-0 has been interferometry with a keV positron beam. Positronium and antimatter interferometry requires addressing specific problems in addition to the scarcity of antiparticles, including the background produced by annihilations, the finite lifetime of Ps (only 142 *ns* for the longest-living ortho-Ps state) and the detection of the interferometric pattern [27].

4. Types of Interferometry

Quantum interferometry can be realized in several different ways. Defining relevant quantities in one of the simplest configurations (two gratings and a detector, Figure 1), one can single out the relevant parameters as:

- The wavelength of the radiation (the de Broglie wavelength of the particle) λ.
- The periodicity of the grating used to evidence the diffraction/interference effect d.
- The longitudinal scale L that is related to the integrating distance or to the observation distance.

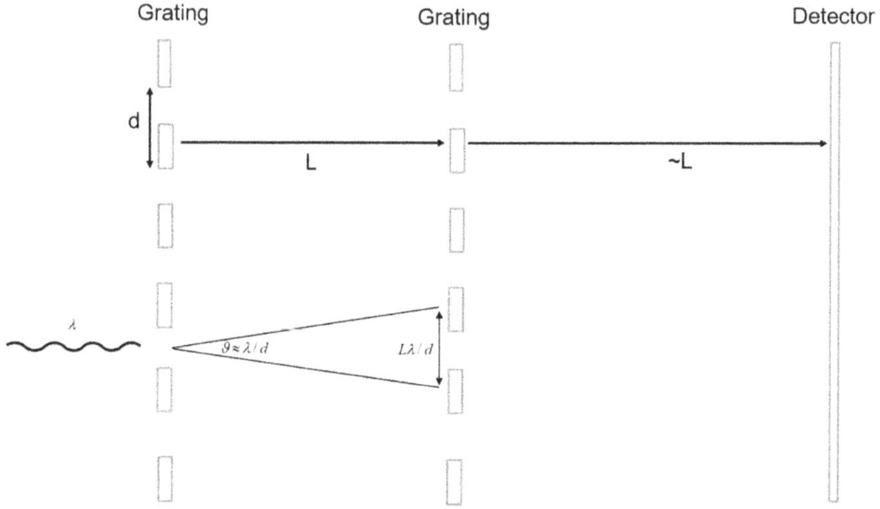

Figure 1. The relevant quantities in an interferometric measurement with gratings are the wavelength of the particle/radiation, the period of the gratings d and the distance scale between gratings *L*. A detector might be located at some distance (similar to *L*) from the second grating. The coherently illuminated area from the first to the second grating is also shown.

The relations which are considered always valid are:

$$\lambda << d \qquad \lambda < L \tag{1}$$

where the first one is often called the large aperture condition. At this point we can form the quantities:

$$D_T = d/\lambda >> 1 \qquad D_L = L/d \tag{2}$$

With respect to λ, D_T is a measure of the dominance of the transverse scale, while D_L indicates the level of longitudinal dominance. If, D_T is big in such a way as to also predominate over D_L then

$$D_T > D_L \rightarrow L < d^2/\lambda \tag{3a}$$

If, on the other hand, it is D_L which dominates D_T, then one has

$$D_T < D_L \rightarrow L > d^2/\lambda \tag{3b}$$

Condition (3a) indicates the so-called moiré regime, or near field interferometry. Under these conditions, the wave-like nature of the particle is not yet evident and the regime is a corpuscular or ballistic one, which is basically the classical physics case. One can have a better appreciation of this when considering a setup like the one in Figure 1. Because of diffraction, for a single slit on grating A the coherence area on B will be $L\lambda/d$. If (3a) holds, then the moiré condition reads $L\lambda/d < d$ and the wave-like nature of the particle will not manifest itself. This regime, more than interferometry, should be more aptly called deflectrometry.

The quantity:

$$L_T = \frac{d^2}{\lambda} \tag{4}$$

is called the Talbot length and is a characteristic of both the grating period and the wavelength under consideration. Three regimes can then be defined whenever (1) is satisfied:

- $L << L_T$: moiré regime, where particles behave like classical bullets (deflectometry).
- $L \approx L_T$: Talbot–Lau regime, where particles start to show interference.
- $L >> L_T$: Fraunhofer regime, where the usual far-field approximation holds.

The Talbot (called the Talbot–Lau regime when multi-slit gratings are used) is an "intermediate field" situation, where the second order term of the development in the Kirchoff integral expansion of wave optics is kept; by contrast, only the first order is considered in the Fraunhofer mode.

The moiré and Talbot regimes have in common the repetition of the produced pattern at integer (and, less evidently, fractional) multiples of the Talbot length. In spite of this numerical similarity, a purely projective effect is at work in the moiré case, while Talbot-mode repetitions are due to a quantum mechanical effect. In other words, the periodicity of the repetition patterns has a purely geometrical origin in the moiré case, while also having a dependence on λ for the Talbot; in this latter case, a change in energy of the particles would also change the longitudinal position of the maxima of the interference pattern.

The Talbot and Fraunhofer configurations both feature the wavelength quantum mechanical dependence of the interference pattern. However, the Talbot case strictly requires the monochromaticity of the beam (and the energy will dictate the position of the repetition pattern). The Fraunhofer case has much less sensitivity to energy so that, when the interference pattern is established, it will always be present at any distance, provided the far-field condition $L >> L_T$ is satisfied. However, the Fraunhofer interference will require a good initial collimation of the beam.

In order to tackle the problem of antimatter interferometry, the positron or the antiproton are the simplest particles of choice. Positron sources are available at linear accelerator (LINAC) machines or by exploiting radioactive sources such as the β^+ emitter ^{22}Na source. Antiprotons are available at particle accelerators since they will need to be produced at very high energy. The Antiproton Decelerator at CERN is the only machine dedicated to the production of antiprotons at the *MeV* scales or below that can prove adequate for interferometry.

5. The Experiment

The QUPLAS-0 experiment, which I will discuss here, is the first stage of the QUPLAS (quantum interferometry and gravitation with positronium) program and consisted in producing the first interferometric pattern with an antiparticle: the positron. For this particular task, a ^{22}Na radioactive source followed by a beam line, an interferometer and a nuclear emulsion detector were used.

For this case, moiré and Talbot configurations are interesting, because of the large momentum acceptance of these configurations. The Fraunhofer requirement of a good beam collimation in fact typically implies a heavy loss of statistics. In addition, the Talbot configuration should be preferred over the moiré in order to put in evidence the quantum mechanical origin of the effect.

After a careful study, the Talbot–Lau setup was considered to be the most promising for the task [28]. One of the main reasons for this is the inevitably poor level of coherence of the beam as well as the need to produce an interference pattern of a minimum periodicity of a few microns to make detection feasible. With reference to Figure 2, the first grating periodicity was $d_1 = 1.2$ μm and the second was $d_2 = 1$ μm, with 50% open fraction in both cases. The gratings and the detector were positioned such that $L_1 \cong 12$ cm and $L_2 \cong 60$ cm (or $L_2 = 5L_1$). This is an example of the so-called Talbot–Lau asymmetric magnifying configuration [25] with a magnification factor of 5, so that the periodicity to be detected at the detector position is of $\cong 5d_1 = 6$ μm. The pattern is therefore detectable by the nuclear emulsion, which has a resolution of about 1 μm [29].

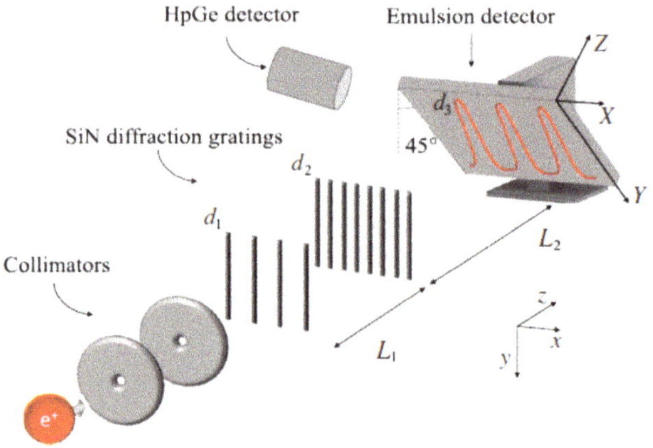

Figure 2. Scheme of the QUPLAS-0 detector experimental configuration. The collimated positron beam propagates through the interferometer, consisting of two gratings and the emulsion detector (tilted by 45°, see text). The interference pattern is collected in the emulsion. A Ge detector is used to monitor the positron beamline through the 511-*keV* gammas generated by positron annihilation.

The configuration of the interferometer and detector system was such as to be resonant at the energy of 14 keV according to the equation

$$\frac{L_1}{L_2} = \frac{d_1}{d_2} - 1 \tag{5}$$

which implicitly contains the Talbot length and the wavelength of the particle by means of (4).

In the final configuration in Figure 2, the emulsion detector was tilted by 45 degrees; this was due to the uncertainty on the longitudinal location of the Talbot revival which is affected by several uncertainties on grating parameters and misalignments (see discussion in [30]).

The experiment made use of the positron beam of the L-NESS Laboratory of the Politecnico di Milano in Como (Italy). The beam had an intensity of about 8×10^3 e^+/s and an angular divergence of a few *mrad*. The positron source is followed by a tungsten (100) moderator and an electrostatic beamline, so that its energy can be tuned between a few keV and 20 keV (with a resolution better than 1%) while maintaining a beam spot of about 2 mm.

6. Results

The QUPLAS-0 data taking took place in 2018, and consisted in a series of exposures of emulsions to the L-NESS positron beam at different energies. After the analysis, the resulting patterns on the detector were studied at 8,9,11 and 14 keV (Figure 3).

In order to investigate the origin of the signal, one has to study the behavior of the visibility (or contrast $C = (I_{max} - I_{min})/(I_{max} + I_{min})$ as a function of the energy, which corresponds to changing the wavelength of the positrons. The result of such a study for energies 8, 9, 11, 14, 16 *keV* is shown in Figure 4. It clearly indicates the quantum mechanical origin of the effect which is energy dependent. By contrast, in the moiré regime no such behavior is expected since the particles would behave classically.

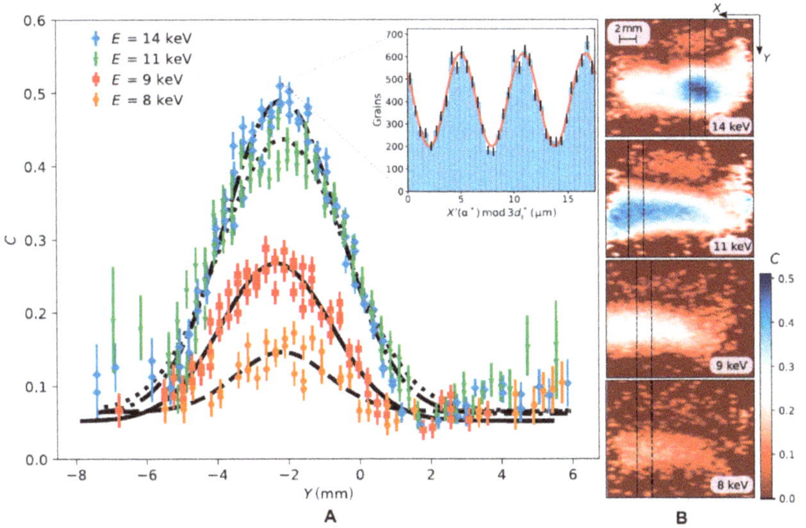

Figure 3. The contrast ($I_{maz} - I_{min}$)/($I_{max} + I_{min}$) is shown on the left (**A**) as a function of the longitudinal coordinate y. It is maximum for the resonant energy of 14 *keV* for which the actual interference pattern is shown in the insert. Other energies are visible albeit with a reduced contrast. On the right (**B**), the transverse position of the interference patterns on the emulsion is shown.

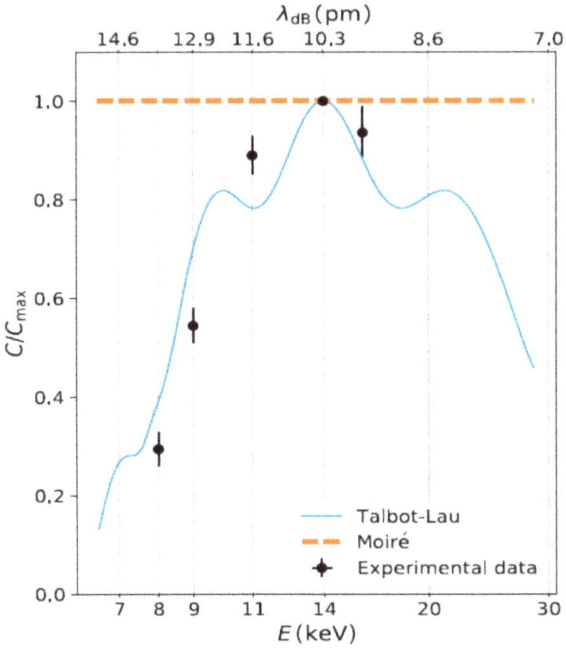

Figure 4. Visibility of the Talbot–Lau interference pattern as a function of energy (wavelength) in QUPLAS-0. The dependence on *E* is the smoking-gun proof of the quantum mechanical origin of the effect. The classical moiré effect (orange dashed line) would in fact have been achromatic.

The result is the first demonstration of antimatter interferometry. In addition, since the flux of particles is at most ~10^4 per second, generated by the time-incoherent ^{22}Na source, and the transit time through the interferometer is 10^{-8} s, this turns out to be a single-particle experiment, being therefore the antimatter version of the celebrated Merli–Missiroli–Pozzi single electron result [11].

7. Conclusions

Quantum interferometry of antimatter has been made for the first time by means of Talbot–Lau interferometry on positrons. This is also the second demonstration ever of single-particle interference obtained with an elementary constituent of the standard model.

Funding: The QUPLAS project is funded by the Politecnico di Milano (Italy) and by the Committee III of the Italian Istituto Nazionale di Fisica Nucleare (under the AEgIS program).

Acknowledgments: The credit for this work goes to the entire QUPLAS group that made possible the series of experiments necessary to reach these goals: S. Aghion, A. Ariga, T. Ariga, M. Bollani, F. Castelli, A. Ereditato, R. Ferragut, M. Leone, M. Lodari, G. Maero, S. Olivares, C. Pistillo, M. Romé, S. Sala, P. Scampoli, and S. Siccardi.

Conflicts of Interest: The author declares no conflict of interest.

References

1. De Broglie, L. Waves and quanta. *Nature* **1923**, *112*, 140. [CrossRef]
2. Davisson, C.J.; Germer, L.H. Reflection of electrons by a crystal of nickel. *Proc. Natl. Acad. Sci. USA* **1928**, *14*, 317. [CrossRef] [PubMed]
3. Thomson, G.P.; Reid, A. Diffraction of cathode rays by a thin film. *Nature* **1927**, *119*, 890. [CrossRef]
4. Rauch, H.; Treimer, W.; Bonse, U. Test of a single crystal neutron interferometer. *Phys. Lett. A* **1974**, *47*, 369. [CrossRef]
5. Overhauser, A.V.; Colella, R. Experimental Test of Gravitationally Induced Quantum Interference. *Phys. Rev. Lett.* **1974**, *33*, 1237. [CrossRef]
6. Colella, R.; Overhauser, A.V.; Werner, S.A. Observation of Gravitationally Induced Quantum Interference. *Phys. Rev. Lett.* **1975**, *34*, 1472. [CrossRef]
7. Zeilinger, A.; Gaehler, R.; Shull, C.G.; Treimer, W.; Mampe, W. Single- and double-slit diffraction of neutrons. *Rev. Mod. Phys.* **1988**, *60*, 106. [CrossRef]
8. Chapman, M.S.; Ekstrom, C.R.; Hammond, T.D.; Rubenstein, R.A.; Schmiedmayer, J.; Wehinger, S.; Pritchard, D.E. Optics and Interferometry with Na_2 Molecules. *Phys. Rev. Lett.* **1995**, *74*, 4783. [CrossRef]
9. Arndt, M.; Nairz, O.; Vos-Andreae, J.; Keller, C.; van der Zouw, G.; Zeilinger, A. Wave-particle duality of C_{60} molecules. *Nature* **1999**, *401*, 680. [CrossRef]
10. Feynman, R. *Feynman Lectures on Physics*; Feynman, R.P., Leighton, R.B., Sands, M., Eds.; Addison-Wesley: Reading, MA, USA, 1965; Volume 3.
11. Merli, P.G.; Missiroli, G.F.; Pozzi, G. On the statistical aspect of electron interference phenomena. *Am. J. Phys.* **1976**, *44*, 306. [CrossRef]
12. Frabboni, S.; Gabrielli, A.; Gazzadi, G.C.; Giorgi, F.M.; Matteucci, G.; Pozzi, G.; Semprini Cesari, N.; Villa, M.; Zoccoli, A. The Young-Feynman two-slit experiment with single electrons: Build-up of the interference pattern and arrival-time distribution using a fast-readout pixel detector. *Ultramicroscopy* **2012**, *116*, 73. [CrossRef]
13. Rosenberg, I.J.; Weiss, A.H.; Canter, K.F. Low-Energy Positron Diffraction from a Cu(111) Surface. *Phys. Rev. Lett.* **1980**, *44*, 1139. [CrossRef]
14. Dirac, P.A.M. The Quantum Theory of the Electron. *Proc. R. Soc. Lond.* **1928**, *A117*, 610. [CrossRef]
15. Anderson, C.D. The Apparent Existence of Easily Deflectable Positives. *Science* **1932**, *76*, 238. [CrossRef]
16. Lueders, G. Proof of the TCP theorem. *Ann. Phys.* **1957**, *2*, 1. [CrossRef]
17. Amoretti, M.; Amsler, C.; Bonomi, G.; Boutcha, A.; Bowe, P.; Carraro, C.; Cesar, C.L.; Charlton, M.; Collier, M.J.T.; Doser, M.; et al. Production and detection of cold antihydrogen atoms. *Nature* **2002**, *419*, 456. [CrossRef]

18. Gabrielse, G.; Bowden, N.S.; Oxley, P.; Speck, A.; Storry, C.H.; Tan, J.N.; Wessels, M.; Grzonka, D.; Oelert, W.; Schepers, G.; et al. Background-Free Observation of Cold Antihydrogen with Field-Ionization Analysis of Its States. *Phys. Rev. Lett.* **2002**, *89*, 213401. [CrossRef] [PubMed]
19. Andresen, G.B.; Ashkezari, M.D.; Baquero-Ruiz, M.; Bertsche, W.; Bowe, P.D.; Butler, E.; Cesar, C.L.; Charlton, M.; Deller, A.; Eriksson, S.; et al. Confinement of antihydrogen for 1,000 seconds. *Nat. Phys.* **2011**, *7*, 558.
20. Kuroda, N.; Ulmer, S.; Murtagh, D.J.; Van Gorp, S.; Nagata, Y.; Diermaier, M.; Federmann, S.; Leali, M.; Malbrunot, C.; Mascagna, V.; et al. A source of antihydrogen for in-flight hyperfine spectroscopy. *Nat. Commun.* **2014**, *5*, 3089. [CrossRef]
21. Deutsch, M. Evidence for the Formation of Positronium in Gases. *Phys. Rev.* **1951**, *82*, 455. [CrossRef]
22. Cassidy, D.B. Experimental progress in positronium laser physics. *Eur. Phys. J. D* **2018**, *72*, 53.
23. Dolgov, A.P. Baryogenesis, 30 years after. *Surv. High Energy Phys.* **1998**, *13*, 83. [CrossRef]
24. Sakharov, A.D. Violation of CP Invariance, C asymmetry, and baryon asymmetry of the universe. *J. Exp. Theory Phys. Lett.* **1967**, *5*, 24.
25. Farrar, G.R.; Shaposhnikov, M.E. Baryon asymmetry of the Universe in the minimal standard model. *Phys. Rev. Lett.* **1993**, *70*, 2833. [CrossRef]
26. Oberthaler, M.K. Anti-matter wave interferometry with positronium. *Nucl. Instr. Methods B* **2002**, *192*, 129. [CrossRef]
27. Sala, S.; Castelli, F.; Giammarchi, M.; Siccardi, S.; Olivares, S. Matter-wave interferometry: towards antimatter interferometers. *J. Phys. B* **2015**, *48*, 195002. [CrossRef]
28. Sala, S.; Giammarchi, M.; Olivares, S. Asymmetric Talbot-Lau interferometry for inertial sensing. *Phys. Rev. A* **2016**, *94*, 033625. [CrossRef]
29. Amsler, C.; Ariga, A.; Ariga, T.; Braccini, S.; Canali, C.; Ereditato, A.; Kawada, J.; Kimura, M.; Kreslo, I.; Pistillo, C.; et al. A new application of emulsions to measure the gravitational force on antihydrogen. *J. Instrum.* **2013**, *8*, P02015. [CrossRef]
30. Sala, S.; Ariga, A.; Ereditato, A.; Ferragut, R.; Giammarchi, M.; Leone, M.; Pistillo, C.; Scampoli, P. First demonstration of antimatter wave interferometry. *Sci. Adv.* **2019**, *5*, eaav7610. [CrossRef]

symmetry

MDPI

Article

Lorentz-Violating Matter-Gravity Couplings in Small-Eccentricity Binary Pulsars

Lijing Shao [1,2]

[1] Kavli Institute for Astronomy and Astrophysics, Peking University, Beijing 100871, China; lshao@pku.edu.cn
[2] Max-Planck-Institut für Radioastronomie, Auf dem Hügel 69, D-53121 Bonn, Germany

Received: 15 August 2019; Accepted: 28 August 2019; Published: 2 September 2019

Abstract: Lorentz symmetry is an important concept in modern physics. Precision pulsar timing was used to put tight constraints on the coefficients for Lorentz violation in the pure-gravity sector of the Standard-Model Extension (SME). We extend the analysis to Lorentz-violating matter-gravity couplings, utilizing three small-eccentricity relativistic neutron star (NS)—white dwarf (WD) binaries. We obtain compelling limits on various SME coefficients related to the neutron, the proton, and the electron. These results are complementary to limits obtained from lunar laser ranging and clock experiments.

Keywords: pulsar timing; Standard-Model Extension; binary pulsars

1. Introduction

The theory of general relativity (GR) and the Standard Model (SM) of particle physics represent our contemporary condensed wisdom in the search of fundamental laws in physics. Nevertheless, there exist various motivations to look for new physics. Among them, the possibility of Lorentz violation is a well developed concept [1]. Lorentz violation could be resulted from a deep underlying theory of quantum gravity [2]. At low energy, it is believed to be described by an effective field theory (EFT). An EFT framework, the so-called Standard-Model Extension (SME), systematically incorporates all Lorentz-covariant, gauge-invariant, energy-momentum-conserving operators that are associated with GR and SM fields [3–5]. Field operators are sorted according to their mass dimension, and, for some certain species, operators of arbitrary mass dimensions are classified [6–9].

The SME is supposed to be an effectively low-energy theory for the quantum gravity, thus, the gravitational aspect of the SME is of particular interest. Kostelecký [5] presented the general structure of the SME when the curved spacetime is considered. Bailey and Kostelecký [10] worked out different kinds of observational phenomena associated with the minimal operators in the pure-gravity sector of the SME, whose mass dimension $d \leq 4$. After that, Kostelecký and Tasson [11] investigated in great detail the theoretical aspects of the matter-gravity couplings, whose mass dimension $d \leq 4$. Phenomenological aspects and relevant experiments are identified. Moreover, the nonminimal SME with gravitational operators, whose mass dimension $d > 4$, was studied and gained global interests during the past few years [12–14].

Due to the advances on the theoretical side [5,10–12], phenomenological and experimental studies of the gravitational SME became a hot topic [15–18]. Hees et al. [19] have a comprehensive summary on this topic—see also the *Data Tables for Lorentz and CPT Violation*, compiled by Kostelecký and Russell [20]. In the pure-gravity sector, binary pulsars turn out to be among the best experiments in constraining (i) the $d \leq 4$ minimal Lorentz-violating operators [21,22]; (ii) dimension-5 CPT-violating operators [23]; as well as (iii) dimension-8 cubic-in-the-Riemannian-tensor operators, which are related to the leading-order violation of the gravitational weak equivalence principle [24]. In a closely related metric-based framework, the so-called parameterized post-Newtonian formalism [25,26], binary pulsars similarly outperform many Solar-system-based experiments [27–30].

In this work, we investigate the matter-gravity couplings in the SME and their signals in binary pulsars [11,31]. In particular, we use small-eccentricity binary pulsars—PSRs J0348+0432 [32], J0751+1807 [33], and J1738+0333 [34]—to put stringent constraints on various matter-gravity coupling coefficients. The limits are compelling, and complementary to other experiments. They contribute to the research field on the experimental examination of the SME.

The paper is organized as follows. In the next section, we review the matter-gravity couplings in the SME [11]. Then, in Section 3, the orbital dynamics for a binary pulsar [31] are provided. In particular, the secular change of the eccentricity vector (decomposed into the two Laplace–Lagrange parameters [35]), and the secular change of the pulsar's projected semimajor axis are discussed. Constraints on the matter-gravity coupling coefficients are given in Section 4. The last section discusses constraints from other experiments, the strong-field aspects of pulsars, and the prospects in improving the limits on the Lorentz-violating matter-gravity couplings.

2. Matter-Gravity Couplings in the SME

In order to incorporate fermion-gravity couplings, we use the vierbein formalism [5]. In the SME, the action for a massive Dirac fermion ψ reads [11]

$$S_\psi = \int e\left(\frac{1}{2} i e^\mu{}_a \overline{\psi} \Gamma^a \overleftrightarrow{D}_\mu \psi - \overline{\psi} M \psi\right) \mathrm{d}^4 x \,, \tag{1}$$

where, for spin-independent cases,

$$\Gamma^a \equiv \gamma^a - c_{\mu\nu} e^{\nu a} e^\mu{}_b \gamma^b - e_\mu e^{\mu a} \,, \tag{2}$$

$$M \equiv m + a_\mu e^\mu{}_a \gamma^a \,. \tag{3}$$

Here, $e_\mu{}^a$ is the vierbein with e as its determinant; m is the mass of the fermion; γ^a is the Dirac matrix; a_μ, $c_{\mu\nu}$, and e_μ are species-dependent, spin-independent coefficient fields for Lorentz violation (see Equations (7) and (8) in [11] for spin-dependent terms).

While being kept to the leading order, a field redefinition via a position-dependent component mixing in the spinor space can be used to show that the CPT-odd coefficients a_μ and e_μ always appear in the combination [11]

$$\left(a_{\mathrm{eff}}\right)_\mu \equiv a_\mu - m e_\mu \,. \tag{4}$$

Therefore, we shall consider only $\left(a_{\mathrm{eff}}\right)_\mu$ and $c_{\mu\nu}$ in the following.

At leading order, the point-particle action is [11],

$$S_u = \int \mathrm{d}\lambda \left[-m\sqrt{-\left(g_{\mu\nu} + 2c_{\mu\nu}\right) u^\mu u^\nu} - \left(a_{\mathrm{eff}}\right)_\mu u^\mu\right] , \tag{5}$$

where $u^\mu \equiv \mathrm{d}x^\mu / \mathrm{d}\lambda$. For a macroscopic composite object, the action Equation (5) is still applicable with the replacements [11],

$$m \to \sum_w N^w m^w \,, \tag{6}$$

$$c_{\mu\nu} \to \frac{\sum_w N^w m^w \left(c^w\right)_{\mu\nu}}{\sum_w N^w m^w} \,, \tag{7}$$

$$\left(a_{\mathrm{eff}}\right)_\mu \to \sum_w N^w \left(a^w_{\mathrm{eff}}\right)_\mu \,, \tag{8}$$

where w denotes the particle species and N^w is the number of particles of type w. We have neglected the contribution from binding energies which could be at most $\sim 20\%$ for neutron stars (NSs), unless some unknown nonperturbative effects take place (see discussions in Section 5) [30]. In general, the role of binding energy could further aid the analysis of signals for Lorentz violation, see Section VI B in [11] for more details. Hereafter, for simplicity we only consider three types of fermions—(i) the electron $w = e$, (ii) the proton $w = p$, and (iii) the neutron $w = n$. In Table 1, we list the estimated composition of these three species for NSs and white dwarfs (WDs), and their corresponding composite coefficient fields for Lorentz violation.

Table 1. Estimated composition for neutron stars (NSs) and white dwarfs (WDs). Composite coefficient fields for Lorentz violation are estimated according to Equations (6)–(8). In the table, $M_{\rm NS}$ and $M_{\rm WD}$ are the masses for NS and WD, respectively, and m^n ($\simeq m^p$) is the mass for a neutron (proton) particle. We define $N_{\rm NS} \equiv M_{\rm NS}/m^n$ and $N_{\rm WD} \equiv M_{\rm WD}/m^n$.

	Neutron Stars	White Dwarfs
Electron number, N^e	~ 0	$\frac{1}{2}N_{\rm WD}$
Proton number, N^p	~ 0	$\frac{1}{2}N_{\rm WD}$
Neutron number, N^n	$N_{\rm NS}$	$\frac{1}{2}N_{\rm WD}$
Composite m	$M_{\rm NS}$	$M_{\rm WD}$
Composite $c_{\mu\nu}$	$c^n_{\mu\nu}$	$\frac{1}{2}\left(c^n_{\mu\nu} + c^p_{\mu\nu} + 0.0005\, c^e_{\mu\nu}\right)$
Composite $\left(a_{\rm eff}\right)_\mu$	$N_{\rm NS}\left(a^n_{\rm eff}\right)_\mu$	$\frac{1}{2}N_{\rm WD}\left[\left(a^n_{\rm eff}\right)_\mu + \left(a^p_{\rm eff}\right)_\mu + \left(a^e_{\rm eff}\right)_\mu\right]$

In general, the coefficient fields, $\left(a_{\rm eff}\right)_\mu$ and $c_{\mu\nu}$, are dynamical fields. In the Riemann–Cartan spacetime, the Lorentz violation often needs to be *spontaneous* [36], instead of *explicit* [5]. The coefficient fields obtain their vacuum expectation values via the Higgs-like spontaneous symmetry breaking mechanism. We denote the vacuum expectation values of $\left(a_{\rm eff}\right)_\mu$ and $c_{\mu\nu}$ as $\left(\bar{a}_{\rm eff}\right)_\mu$ and $\bar{c}_{\mu\nu}$, respectively. The barred quantities are also known as the *coefficients for Lorentz violation* [20]. In asymptotically inertial Cartesian coordinates, they are assumed to be small and satisfy [11]

$$\partial_\alpha \left(\bar{a}_{\rm eff}\right)_\mu = 0\,, \tag{9}$$

$$\partial_\alpha \bar{c}_{\mu\nu} = 0\,. \tag{10}$$

The coefficients for Lorentz violation, $\left(\bar{a}_{\rm eff}\right)_\mu$ and $\bar{c}_{\mu\nu}$ [20], are the quantities that we want to investigate with pulsar timing experiments [37,38] in this work.

3. Binary Pulsars with Lorentz-Violating Matter-Gravity Couplings

Jennings et al. [31] worked out the osculating elements for a binary system, composed of masses M_1 and M_2, in the presence of the Lorentz-violating matter-gravity couplings. We consistently use the subscript "1" to denote the pulsar; and use the subscript "2" to denote the companion which is a WD in our study. We define $q \equiv M_1/M_2$ and $M \equiv M_1 + M_2$. To simplify some expressions, we also define $X \equiv M_1/M = q/\left(1+q\right)$, then, $M_2/M = 1 - X = 1/\left(1+q\right)$.

Neglecting the finite-size effects, the Newtonian relative acceleration for a binary is $\boldsymbol{a}_{\rm N} = -GM_1M_2/r^2\hat{\boldsymbol{r}}$, where $\boldsymbol{r}$ is the relative separation and $\hat{\boldsymbol{r}} \equiv \boldsymbol{r}/r$. In the Newtonian gravity, a two-body system with a negative total orbital energy forms an elliptical orbit. An elliptical orbit in the celestial mechanics is usually described by six orbital elements—(i) the semimajor axis a; (ii) the orbital eccentricity e; (iii) the epoch of periastron passage T_0; (iv) the inclination of orbit i; (v) the longitude of periastron ω; and (vi) the longitude of ascending node Ω. The last three angles are illustrated in Figure 1.

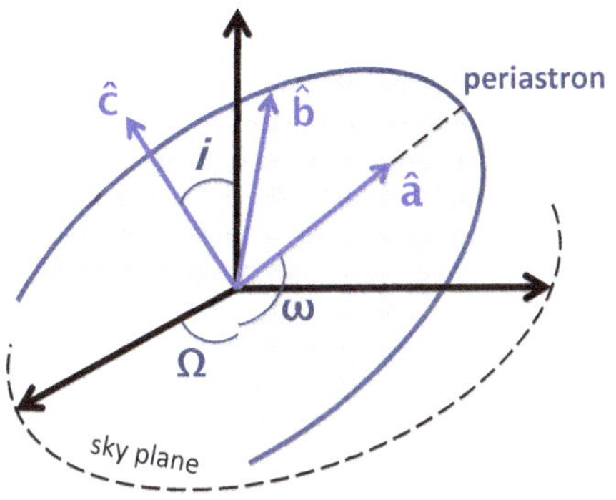

Figure 1. Pulsar orbit and the coordinate system $\left(\hat{\mathfrak{a}}, \hat{\mathfrak{b}}, \hat{\mathfrak{c}}\right)$ [10,22,23].

When there is a perturbing acceleration to $\boldsymbol{a}_{\rm N}$, say, $\delta\boldsymbol{a}$, the orbit is changed perturbatively. In the osculating-element approach, one assumes that at any instant moment, the orbit is still an ellipse, but the six orbital elements become functions of the time t [39]. The time derivatives of these six functions are derived from the extra acceleration $\delta\boldsymbol{a}$ [39]. In the current case, after averaging over an orbital period P_b, the secular changes read [31]

$$\left\langle \frac{\mathrm{d}a}{\mathrm{d}t} \right\rangle = 0, \tag{11}$$

$$\left\langle \frac{\mathrm{d}e}{\mathrm{d}t} \right\rangle = \frac{n_b}{M}\gamma\left(\frac{e^2-2\varepsilon}{e^3}A_{\hat{\mathfrak{a}}\hat{\mathfrak{b}}} + \frac{n_b a\varepsilon}{e^2}B_{\hat{\mathfrak{a}}}\right), \tag{12}$$

$$\left\langle \frac{\mathrm{d}i}{\mathrm{d}t} \right\rangle = \frac{n_b}{M\gamma}\left(\frac{\varepsilon}{e^2}A_{\hat{\mathfrak{a}}\hat{\mathfrak{c}}}\cos\omega - \frac{e^2-\varepsilon}{e^2}A_{\hat{\mathfrak{b}}\hat{\mathfrak{c}}}\sin\omega - \frac{n_b\varepsilon a}{e}B_{\hat{\mathfrak{c}}}\sin\omega\right), \tag{13}$$

$$\begin{aligned}\left\langle \frac{\mathrm{d}\omega}{\mathrm{d}t} \right\rangle =& -\frac{n_b}{M\gamma\tan i}\left(\frac{\varepsilon}{e^2}A_{\hat{\mathfrak{a}}\hat{\mathfrak{c}}}\sin\omega + \frac{e^2-\varepsilon}{e^2}A_{\hat{\mathfrak{b}}\hat{\mathfrak{c}}}\cos\omega + \frac{n_b\varepsilon a}{e}B_{\hat{\mathfrak{c}}}\cos\omega\right) \\ &+ \frac{n_b}{M}\left[\frac{e^2-2\varepsilon}{2e^4}\left(A_{\hat{\mathfrak{b}}\hat{\mathfrak{b}}} - A_{\hat{\mathfrak{a}}\hat{\mathfrak{a}}}\right) + \frac{n_b a(1-\gamma)}{e^3}B_{\hat{\mathfrak{b}}}\right],\end{aligned} \tag{14}$$

where we have defined $\gamma \equiv \sqrt{1-e^2}$, $\varepsilon \equiv 1-\gamma = 1-\sqrt{1-e^2}$, and $n_b \equiv 2\pi/P_b$. From Equation (11), we can see that the energy of the orbit is conserved at leading order, which is compatible with the action formulation of the system in the absence of gravitational waves. The expressions for $\langle \mathrm{d}\Omega/\mathrm{d}t \rangle$ and $\langle \mathrm{d}T_0/\mathrm{d}t \rangle$ are not important in the present context, and thus not shown. The 3-vector B_j and the 3×3 tensor A_{jl} are defined as [31],

$$A_{jl} = \sum_w 2n_7^w m^w \overline{c}^w_{(jl)}, \tag{15}$$

$$B_j = -\sum_w 2\left[n_2^w\left(\overline{a}^w_{\rm eff}\right)_j + \left(n_6^w - 2n_8^w\right)m^w\overline{c}^w_{(0j)}\right], \tag{16}$$

where n_i^w $(i = 1, \cdots, 8)$ are defined in Equation (9) of [31], and their approximated values for NS–NS and NS–WD binaries are given in Table 2 for convenience.

Table 2. Expressions of n_i^w/N ($i = 1, \cdots, 8$; $w \in \{n, p, e\}$) for NS–NS and NS–WD systems (see Equation (9) in [31]), where $N \equiv N_1 + N_2 \simeq M/m^n$. Results in Table 1 are adopted for the calculation here.

	Neutron Star–Neutron Star			Neutron Star–White Dwarf		
	n	p	e	n	p	e
n_1^w/N	1	0	0	$\frac{1}{2}(1+X)$	$\frac{1}{2}(1-X)$	$\frac{1}{2}(1-X)$
n_2^w/N	$2X-1$	0	0	$\frac{1}{2}(3X-1)$	$-\frac{1}{2}(1-X)$	$-\frac{1}{2}(1-X)$
n_3^w/N	2	0	0	$\frac{3}{2}$	$\frac{1}{2}$	$\frac{1}{2}$
n_4^w/N	0	0	0	$-\frac{1}{2}$	$\frac{1}{2}$	$\frac{1}{2}$
n_5^w/N	$2X(1-X)$	0	0	$\frac{3}{2}X(1-X)$	$\frac{1}{2}X(1-X)$	$\frac{1}{2}X(1-X)$
n_6^w/N	0	0	0	$-\frac{1}{2}X(1-X)$	$\frac{1}{2}X(1-X)$	$\frac{1}{2}X(1-X)$
n_7^w/N	1	0	0	$1-\frac{1}{2}X$	$\frac{1}{2}X$	$\frac{1}{2}X$
n_8^w/N	$1-2X$	0	0	$\frac{1}{2}X^2-2X+1$	$-\frac{1}{2}X^2$	$-\frac{1}{2}X^2$

In the above two equations, only n_i^w with $i = 2, 6, 7, 8$ are relevant. Using the results in Table 2, we have

$$
\begin{aligned}
\frac{A_{jl}}{M} =& (2-X)\bar{c}^n_{(jl)} + X\left[\bar{c}^p_{(jl)} + 0.0005\bar{c}^e_{(jl)}\right], \qquad (17)\\
\frac{B_j}{M} =& \frac{1-X}{m^n}\left[\left(\bar{a}^p_{\text{eff}}\right)_j + \left(\bar{a}^e_{\text{eff}}\right)_j\right] + \frac{1-3X}{m^n}\left(\bar{a}^n_{\text{eff}}\right)_j \\
&+ \left(X^2 - 7X + 4\right)\bar{c}^n_{(0j)} - X(1+X)\left[\bar{c}^p_{(0j)} + 0.0005\bar{c}^e_{(0j)}\right]. \qquad (18)
\end{aligned}
$$

We can easily obtain the following conclusion from the above two equations. (I) The sensitivity to $\bar{c}^e_{(jl)}$ and $\bar{c}^e_{(0j)}$ (compared with $\bar{c}^p_{(jl)}$ and $\bar{c}^p_{(0j)}$, respectively) is suppressed by the mass ratio of the electron to the proton ($m^e/m^p \simeq 0.0005$), while the sensitivity to $\left(\bar{a}^e_{\text{eff}}\right)_j$ (compared with $\left(\bar{a}^p_{\text{eff}}\right)_j$) is not suppressed. (II) We have no sensitivity to $\left(\bar{a}^w_{\text{eff}}\right)_0$ nor $\bar{c}^w_{00}$ ($w \in \{n, p, e\}$) from binary pulsars in this simplified situation. This is similar to the case of $\bar{s}_{00}$ (the time–time component of the Lorentz-violating field $\bar{s}_{\mu\nu}$) in the pure-gravity sector of the SME with dimension 4 operators [10,21], nevertheless, these coefficients can be probed with the help of the "boost effect" introduced by the systematic velocity of the binary ($v^{\text{sys}}/c \sim 10^{-3}$) with respect to the Solar system [22]. We defer the investigation along this line to future studies.

In Equations (11)–(14), B_j and A_{jl} are projected to the coordinate system $\left(\hat{\mathbf{a}}, \hat{\mathbf{b}}, \hat{\mathbf{c}}\right)$ [10,22,23], where $\hat{\mathbf{a}}$ is the unit vector points from the center of binary towards the periastron, $\hat{\mathbf{c}}$ is the unit vector points along the orbital angular momentum, and $\hat{\mathbf{b}} \equiv \hat{\mathbf{c}} \times \hat{\mathbf{a}}$ (see Figure 1).

We are interested in the small-eccentricity binaries. In the limiting case of small eccentricity $e \to 0$, we have

$$
\gamma = 1 - \frac{1}{2}e^2 - \frac{1}{8}e^4 + \mathcal{O}\left(e^6\right), \qquad (19)
$$

$$
\varepsilon = \frac{1}{2}e^2 + \frac{1}{8}e^4 + \mathcal{O}\left(e^6\right). \qquad (20)
$$

Therefore, Equations (12)–(14) are simplified to

$$\left\langle \frac{\mathrm{d}e}{\mathrm{d}t} \right\rangle \simeq \frac{n_b^2 a}{2M} B_{\hat{a}} \,, \tag{21}$$

$$\left\langle \frac{\mathrm{d}i}{\mathrm{d}t} \right\rangle \simeq \frac{n_b}{2M} \left(A_{\hat{a}\hat{c}} \cos\omega - A_{\hat{b}\hat{c}} \sin\omega \right) \,, \tag{22}$$

$$\left\langle \frac{\mathrm{d}\omega}{\mathrm{d}t} \right\rangle \simeq \frac{n_b^2 a}{2eM} B_{\hat{b}} \,. \tag{23}$$

We can convert the derivatives of e, i, and ω into derivatives of the projected semimajor axis of the pulsar orbit x_p, and the Laplace–Lagrange parameters, $\eta \equiv e \sin\omega$ and $\kappa \equiv e\cos\omega$ into

$$\left\langle \frac{\mathrm{d}x_p}{\mathrm{d}t} \right\rangle = \frac{M_2 \cos i}{2M^2} \left(GMn_b \right)^{1/3} \left(A_{\hat{a}\hat{c}} \cos\omega - A_{\hat{b}\hat{c}} \sin\omega \right) \,, \tag{24}$$

$$\left\langle \frac{\mathrm{d}\eta}{\mathrm{d}t} \right\rangle = \frac{n_b}{2M} \left(GMn_b \right)^{1/3} \left(B_{\hat{a}} \sin\omega + B_{\hat{b}} \cos\omega \right) \,, \tag{25}$$

$$\left\langle \frac{\mathrm{d}\kappa}{\mathrm{d}t} \right\rangle = \frac{n_b}{2M} \left(GMn_b \right)^{1/3} \left(B_{\hat{a}} \cos\omega - B_{\hat{b}} \sin\omega \right) \,, \tag{26}$$

where we have used $n_b a = \left(GMn_b \right)^{1/3}$.

4. Bounds on the SME Coefficients

We use the time derivatives of x_p, η, and κ in Equations (24)–(26) to constrain the coefficients for Lorentz violation. It is clear that the more relativistic the binary (namely, the larger n_b), the better the tests. Therefore, we use three well-timed NS–WD binaries whose orbital periods are shorter than half a day [32–34]. Relevant parameters of these binaries are collected in Table 3. Due to the binary interaction and matter exchange in the evolutionary history, these NS–WD binaries have small orbital eccentricity $e \leq 10^{-6}$, thus, Equations (24)–(26) are sufficient to perform the tests.

Table 3. Relevant parameters for PSRs J0348+0432 [32], J0751+1807 [33], and J1738+0333 [34]. Parenthesized numbers represent the 1-σ uncertainty in the last digit(s) quoted. The parameter η is the intrinsic value, after subtraction of the contribution from the Shapiro delay [35]. Masses are derived without using information related to $\langle \mathrm{d}x_p/\mathrm{d}t \rangle$, $\langle \mathrm{d}\eta/\mathrm{d}t \rangle$, nor $\langle \mathrm{d}\kappa/\mathrm{d}t \rangle$ for consistency. For PSRs J0348+0432 and J1738+0333, masses were derived independently of gravity theories [32,34], while for PSR J0751+1807 we have used observed quantities related to the Shapiro delay and orbital decay, assuming the validity of general relativity (GR) [33].

Pulsar	J0348+0432	J0751+1807	J1738+0333
Observational span, $T_{\rm obs}$ (year)	$\sim$3.7	$\sim$17.6	$\sim$10.0
Orbital period, P_b (day)	0.102424062722(7)	0.263144270792(7)	0.3547907398724(13)
Pulsar's projected semimajor axis, x_p (lt-s)	0.14097938(7)	0.3966158(3)	0.343429130(17)
$\eta \equiv e\sin\omega\,(10^{-7})$	19(10)	33(5)	$-1.4(11)$
$\kappa \equiv e\cos\omega\,(10^{-7})$	14(10)	3.8(50)	3.1(11)
Time derivative of x_p, $\dot{x}_p$	–	$(-4.9 \pm 0.9) \times 10^{-15}$	$(0.7 \pm 0.5) \times 10^{-15}$
NS mass, m_1 ($M_\odot$)	2.01(4)	1.64(15)	$1.46^{+0.06}_{-0.05}$
WD mass, m_2 ($M_\odot$)	0.172(3)	0.16(1)	$0.181^{+0.008}_{-0.007}$

From Table 3, we see that the time derivatives of η and κ are not reported in literature, as well as the time derivative of x_p for PSR J0348+0432. The reason is usually the following. In fitting the times of arrival of pulse signals, these quantities would be measured to be consistent with zero if they were included in the timing formalism. To have a simpler timing model, these quantities are considered *unnecessary* for a good fit. Actually, the insignificance of these quantities is consistent with the spirit of our tests to put upper limits on the Lorentz violation. We estimate the upper limits for these quantities using $\dot{X} \sim \sqrt{12}\sigma_X/T_{\rm obs}$ ($X \in \{x_p, \eta, \kappa\}$) [21], where σ_X is the measured uncertainty for the quantity X and $T_{\rm obs}$ is the observational span of the data from where these quantities were derived. The factor "$\sqrt{12}$" approximately takes a linear-in-time evolution of the quantity X into account [21]. It is verified that this approximation works reasonably well [21,23]. For PSRs J0751+1807 and J1738+0333, $\langle {\rm d}x_p/{\rm d}t \rangle$ was measured to be nonzero. As the proper motion of the binary in the sky could contribute to a nonzero $\langle {\rm d}x_p/{\rm d}t \rangle$ for nearby pulsars [37,40], we use the measured value of $\langle {\rm d}x_p/{\rm d}t \rangle$ as an upper limit for the effects from Lorentz violation. For nearby pulsars, the contribution to $\langle {\rm d}x_p/{\rm d}t \rangle$ from the proper motion depends sinusoidally on Ω [37,40]—although Ω is not measured, we do not expect the Nature's conspiracy in assigning certain values of Ω case-by-case to different binary pulsars, in order to hide the Lorentz symmetry breaking. Therefore, we believe the above treatments introduce uncertainties no larger than a multiplicative factor of a few.

In order to use Equations (24)–(26), one also needs the absolute geometry of the orbit to properly project the vector B_j and the tensor A_{jl} onto the coordinate system $\left(\hat{\mathbf{a}}, \hat{\mathbf{b}}, \hat{\mathbf{c}}\right)$. In general, the longitude of the ascending node Ω is not observable in pulsar timing [37]. Nevertheless, the procedure to randomize the value of $\Omega \in [0, 360^\circ)$ and to systematically project vectors and tensors onto $\left(\hat{\mathbf{a}}, \hat{\mathbf{b}}, \hat{\mathbf{c}}\right)$ was worked out in [21]. It was successfully applied to binary pulsars in previous studies [21–24]. Since here (i) we have already introduced an uncertainty with a factor of a few, and (ii) we are interested in the "maximal-reach" limits in absence of the Lorentz violation, we take a simplified approach and treat these projections as $\mathcal{O}(1)$ operators. The "maximal-reach" approach [18] assumes that only one component of Lorentz-violating coefficients is nonzero in a test. We think our approach is reasonable at the stage of setting upper limits to the coefficients for Lorentz violation. Nevertheless, when people start to discover some evidence for the Lorentz violation, it is *absolutely* needed to take into account more sophisticated analysis, for example, to use the 3-D orientation of the orbit (possibly in a probabilistic way with an unknown Ω) as was done in [21–24]. In addition, if one wants to explore the correlation between different coefficients for Lorentz violation, more sophisticated analysis is needed as well. These improvements lay beyond the scope of this work.

In Table 4, we list the "maximal-reach" [18] limits on the coefficients for Lorentz violation with matter-gravity couplings obtained from binary pulsars. As we can see, the best limits on $\bar{c}^w_{jk}$ ($w \in \{n, p, e\}$) come from PSR J1738+0333 due to its very good measurement on the $\dot{x}_p$ [34]. For $\bar{c}^w_{0k}$ and $\left(\bar{a}^w_{\rm eff}\right)_k$, the best limits come from PSR J0751+1807 due to its good measurement of the Lagrange–Laplace parameters [33].

Table 4. "Maximal-reach" limits from binary pulsars on the coefficients for Lorentz violation with matter-gravity couplings, where only one component is assumed to be nonzero at a time. The limits on $\bar{c}^w_{jk}$ ($w \in \{n, p, e\}$) come from $\langle {\rm d}x_p/{\rm d}t \rangle$; while the limits on $\bar{c}^w_{0k}$ and $\left(\bar{a}^w_{\rm eff}\right)_k$ come from $\langle {\rm d}\eta/{\rm d}t \rangle$ or $\langle {\rm d}\kappa/{\rm d}t \rangle$, and only the stronger one is listed in the table. For each row, the strongest limit is shown in boldface.

SME Coefficients	PSR J0348+0432	PSR J0751+1807	PSR J1738+0333
$\bar{c}^n_{jk}$	3×10^{-11}	2×10^{-10}	$\mathbf{1 \times 10^{-11}}$
$\bar{c}^p_{jk}$	4×10^{-11}	2×10^{-10}	$\mathbf{1 \times 10^{-11}}$
$\bar{c}^e_{jk}$	8×10^{-8}	4×10^{-7}	$\mathbf{3 \times 10^{-8}}$
$\bar{c}^n_{0k}$	3×10^{-8}	$\mathbf{1 \times 10^{-8}}$	7×10^{-8}
$\bar{c}^p_{0k}$	2×10^{-8}	$\mathbf{1 \times 10^{-8}}$	6×10^{-8}
$\bar{c}^e_{0k}$	5×10^{-5}	$\mathbf{2 \times 10^{-5}}$	1×10^{-4}
$\left(\bar{a}^n_{\rm eff}\right)_k$	2×10^{-8} GeV	$\mathbf{1 \times 10^{-8}}$ **GeV**	6×10^{-8} GeV
$\left(\bar{a}^p_{\rm eff}\right)_k$	5×10^{-7} GeV	$\mathbf{2 \times 10^{-7}}$ **GeV**	8×10^{-7} GeV
$\left(\bar{a}^e_{\rm eff}\right)_k$	5×10^{-7} GeV	$\mathbf{2 \times 10^{-7}}$ **GeV**	8×10^{-7} GeV

5. Discussion

Besides the streamlined theoretical analysis, the maximal-reach limits in Table 4 are the main results of this paper. As far as we are aware, Altschul [41] was the first to put preliminary limits on the SME neutron-sector coefficients with pulsar rotations. The pure-gravity sector of the SME at different mass dimensions was systematically tested with binary pulsars in [21–24]. Early limits on $(\bar{a}^{w}_{\rm eff})_k$ were given with K/He magnetometer and torsion-strip balance [42,43], but these limits, while constraining different linear combinations of the Lorentz violating coefficients, are rather weak. Later, the maximal-reach limits on $(\bar{a}^{w}_{\rm eff})_k$ were obtained systematically with superconducting gravimeters [44] and lunar laser ranging (LLR) experiments [45]. The former got $(\bar{a}^{w}_{\rm eff})_k \leq \mathcal{O}\left(10^{-5}\,{\rm GeV}\right)$; while the latter got $(\bar{a}^{w}_{\rm eff})_k \leq \mathcal{O}\left(10^{-8}\,{\rm GeV}\right)$. Our best limits from PSR J0751+1807 for the proton and the electron are weaker than the LLR limits, while our limit for the neutron is slightly better. There is also a limit from the observation of gravitational waves, but being weaker than our limits by almost 30 orders of magnitude [46]. The limits on $(\bar{a}^{w}_{\rm eff})_0$ were cast by analyzing nuclear binding energy, Cs interferometer, torsion pendulum, and weak equivalence principle experiments [11,47–49]. The analysis with binary pulsars in this work could not bound these SME coefficients. The limits on $\bar{c}^{w}_{\mu\nu}$ from other experiments (e.g., clock experiments [50]) are much better than the limits from binary pulsars [20]. However, our limits are the best ones from gravitational systems. In a short summary, our limits are compelling, and complementary to limits obtained from other experiments.

In using the SME, we have assumed the validity of the effective field theory (EFT) and the smallness of the Lorentz violation. This is true for most ordinary objects. However, we shall be aware of a caveat for NSs, because of the possible nonperturbative behaviors which might be triggered by their strongly self-gravitating nature [38]. It was shown explicitly that, in a class of scalar-tensor theories, highly nonlinear phenomena are possible within NSs and they may result in large deviations from GR [51,52]. Although the nonperturbative behaviors were constrained tightly with binary pulsars and the binary neutron star inspiral GW170817 [34,53,54], the possibility is not completely ruled out yet [55–57]. With this caveat in mind, conservatively speaking, the tests in this paper are basically testing the strong-field counterparts of the weak-field SME coefficients. Usually, when the strong-field effects are considered, the constraints become even tighter. Therefore, we treat the limits here as conservative ones [30].

The tests of Lorentz violation with binary pulsars improve with a longer baseline for data [21]. Specifically, even *pessimistically* assuming no advance in the quality of binary-pulsar observation for the future, the tests in Equations (24)–(26) improve as $T_{\rm obs}^{-1.5}$, where $T_{\rm obs}$ is the total observational span. In reality, the quality of observation improves rapidly, especially with the newly built and upcoming telescopes, like the Five-hundred-meter Aperture Spherical Telescope (FAST), the MeerKAT telescope, and the Square Kilometre Array (SKA) [58–61]. Therefore, we expect better tests than the $T_{\rm obs}^{-1.5}$ scaling in testing the Lorentz violation in the future.

Funding: This work was supported by the National Natural Science Foundation of China (11975027), the Young Elite Scientists Sponsorship Program by the China Association for Science and Technology (2018QNRC001), and partially supported by the National Natural Science Foundation of China (11721303), the Strategic Priority Research Program of the Chinese Academy of Sciences through the Grant No. XDB23010200, and the European Research Council (ERC) for the ERC Synergy Grant BlackHoleCam under Contract No. 610058.

Acknowledgments: We are grateful to Jay Tasson for the invitation and stimulating discussions. We thank Norbert Wex for carefully reading the manuscript, and Adrien Bourgoin, Zhi Xiao, and Rui Xu for communication.

Conflicts of Interest: The author declares no conflict of interest.

Abbreviations

The following abbreviations are used in this manuscript:

EFT	Effective Field Theory
GR	General Relativity
LLR	Lunar Laser Ranging
NS	Neutron Star
SM	Standard Model
SME	Standard-Model Extension
WD	White Dwarf

References

1. Tasson, J.D. What Do We Know About Lorentz Invariance? *Rept. Prog. Phys.* **2014**, *77*, 062901. [CrossRef] [PubMed]
2. Kostelecký, V.A.; Samuel, S. Spontaneous Breaking of Lorentz Symmetry in String Theory. *Phys. Rev. D* **1989**, *39*, 683. [CrossRef] [PubMed]
3. Colladay, D.; Kostelecký, V.A. CPT violation and the standard model. *Phys. Rev. D* **1997**, *55*, 6760–6774. [CrossRef]
4. Colladay, D.; Kostelecký, V.A. Lorentz violating extension of the standard model. *Phys. Rev. D* **1998**, *58*, 116002. [CrossRef]
5. Kostelecký, V.A. Gravity, Lorentz violation, and the standard model. *Phys. Rev. D* **2004**, *69*, 105009. [CrossRef]
6. Kostelecký, V.A.; Mewes, M. Electrodynamics with Lorentz-violating operators of arbitrary dimension. *Phys. Rev. D* **2009**, *80*, 015020. [CrossRef]
7. Kostelecký, V.A.; Mewes, M. Neutrinos with Lorentz-violating operators of arbitrary dimension. *Phys. Rev. D* **2012**, *85*, 096005. [CrossRef]
8. Kostelecký, A.; Mewes, M. Fermions with Lorentz-violating operators of arbitrary dimension. *Phys. Rev. D* **2013**, *88*, 096006. [CrossRef]
9. Kostelecký, V.A.; Li, Z. Gauge field theories with Lorentz-violating operators of arbitrary dimension. *Phys. Rev. D* **2019**, *99*, 056016. [CrossRef]
10. Bailey, Q.G.; Kostelecký, V.A. Signals for Lorentz violation in post-Newtonian gravity. *Phys. Rev. D* **2006**, *74*, 045001. [CrossRef]
11. Kostelecký, V.A.; Tasson, J.D. Matter-gravity couplings and Lorentz violation. *Phys. Rev. D* **2011**, *83*, 016013. [CrossRef]
12. Bailey, Q.G.; Kostelecký, V.A.; Xu, R. Short-range gravity and Lorentz violation. *Phys. Rev. D* **2015**, *91*, 022006. [CrossRef]
13. Shao, C.G.; Tan, Y.-J.; Tan, W.-H.; Yang, S.-Q.; Luo, J.; Tobar, M.E.; Bailey, Q.G.; Long, J.C.; Weisman, E.; Xu, R.; et al. Combined search for Lorentz violation in short-range gravity. *Phys. Rev. Lett.* **2016**, *117*, 071102. [CrossRef] [PubMed]
14. Kostelecký, V.A.; Mewes, M. Lorentz and Diffeomorphism Violations in Linearized Gravity. *Phys. Lett. B* **2018**, *779*, 136–142. [CrossRef]
15. Tasson, J.D. Gravitational Searches for Lorentz Violation with Matter and Astrophysics. In Proceedings of the 7th Meeting on CPT and Lorentz Symmetry (CPT 16), Bloomington, IN, USA, 20–24 June 2016; pp. 13–16.
16. Bailey, Q.G. Recent Developments in Spacetime-Symmetry tests in Gravity. In Proceedings of the 8th Meeting on CPT and Lorentz Symmetry (CPT'19), Bloomington, IN, USA, 12–16 May 2019.
17. Shao, L. Pulsar Tests of the Gravitational Lorentz Violation. In Proceedings of the 8th Meeting on CPT and Lorentz Symmetry (CPT'19), Bloomington, IN, USA, 12–16 May 2019.
18. Tasson, J.D. Maximal Tests in Minimal Gravity. In Proceedings of the 8th Meeting on CPT and Lorentz Symmetry (CPT'19), Bloomington, IN, USA, 12–16 May 2019.
19. Hees, A.; Bailey, Q.G.; Bourgoin, A.; Bars, H.P.L.; Guerlin, C.; Le Poncin-Lafitte, C. Tests of Lorentz symmetry in the gravitational sector. *Universe* **2016**, *2*, 30. [CrossRef]
20. Kostelecký, V.A.; Russell, N. Data Tables for Lorentz and CPT Violation. *Rev. Mod. Phys.* **2011**, *83*, 11–31. [CrossRef]
21. Shao, L. Tests of local Lorentz invariance violation of gravity in the standard model extension with pulsars. *Phys. Rev. Lett.* **2014**, *112*, 111103. [CrossRef]

22. Shao, L. New pulsar limit on local Lorentz invariance violation of gravity in the standard-model extension. *Phys. Rev. D* **2014**, *90*, 122009. [CrossRef]
23. Shao, L.; Bailey, Q.G. Testing velocity-dependent CPT-violating gravitational forces with radio pulsars. *Phys. Rev. D* **2018**, *98*, 084049. [CrossRef]
24. Shao, L.; Bailey, Q.G. Testing the Gravitational Weak Equivalence Principle in the Standard-Model Extension with Binary Pulsars. *Phys. Rev. D* **2019**, *99*, 084017. [CrossRef]
25. Will, C.M. The Confrontation between General Relativity and Experiment. *Living Rev. Rel.* **2014**, *17*, 4. [CrossRef] [PubMed]
26. Will, C.M. *Theory and Experiment in Gravitational Physics*; Cambridge University Press: Cambridge, UK, 2018.
27. Shao, L.; Wex, N. New tests of local Lorentz invariance of gravity with small-eccentricity binary pulsars. *Class. Quant. Grav.* **2012**, *29*, 215018. [CrossRef]
28. Shao, L.; Caballero, R.N.; Kramer, M.; Wex, N.; Champion, D.J.; Jessner, A. A new limit on local Lorentz invariance violation of gravity from solitary pulsars. *Class. Quant. Grav.* **2013**, *30*, 165019. [CrossRef]
29. Shao, L.; Wex, N. New limits on the violation of local position invariance of gravity. *Class. Quant. Grav.* **2013**, *30*, 165020. [CrossRef]
30. Shao, L.; Wex, N. Tests of gravitational symmetries with radio pulsars. *Sci. China Phys. Mech. Astron.* **2016**, *59*, 699501. [CrossRef]
31. Jennings, R.J.; Tasson, J.D.; Yang, S. Matter-Sector Lorentz Violation in Binary Pulsars. *Phys. Rev. D* **2015**, *92*, 125028. [CrossRef]
32. Antoniadis, J.; Freire, P.C.; Wex, N.; Tauris, T.M.; Lynch, R.S.; van Kerkwijk, M.H.; Kramer, M.; Bassa, C.; Dhillon, V.S.; Driebe, T.; et al. A Massive Pulsar in a Compact Relativistic Binary. *Science* **2013**, *340*, 448. [CrossRef]
33. Desvignes, G.; Caballero, R.N.; Lentati, L.; Verbiest, J.P.W.; Champion, D.J.; Stappers, B.W.; Janssen, G.H.; Lazarus, P.; Osłowski, S.; Babak, S.; et al. High-precision timing of 42 millisecond pulsars with the European Pulsar Timing Array. *Mon. Not. Roy. Astron. Soc.* **2016**, *458*, 3341–3380. [CrossRef]
34. Freire, P.C.C.; Wex, N.; Esposito-Farèse, G.; Verbiest, J.P.W.; Bailes, M.; Jacoby, B.A.; Kramer, M.; Stairs, I.H.; Antoniadis, J.; Janssen, G.H. The relativistic pulsar-white dwarf binary PSR J1738+0333 II. The most stringent test of scalar-tensor gravity. *Mon. Not. Roy. Astron. Soc.* **2012**, *423*, 3328. [CrossRef]
35. Lange, C.; Camilo, F.; Wex, N.; Kramer, M.; Backer, D.C.; Lyne, A.G.; Doroshenko, O. Precision timing measurements of psr j1012+5307. *Mon. Not. Roy. Astron. Soc.* **2001**, *326*, 274. [CrossRef]
36. Bluhm, R.; Bossi, H.; Wen, Y. Gravity with explicit spacetime symmetry breaking and the Standard-Model Extension. *arXiv* **2019**, arXiv:1907.13209.
37. Lorimer, D.R.; Kramer, M. *Handbook of Pulsar Astronomy*; Cambridge University Press: Cambridge, UK, 2005.
38. Wex, N. Testing Relativistic Gravity with Radio Pulsars. In *Frontiers in Relativistic Celestial Mechanics: Applications and Experiments*; Kopeikin, S.M., Ed.; Walter de Gruyter GmbH: Berlin, Germany; Boston, MA, USA, 2014; Volume 2, p. 39.
39. Poisson, E.; Will, C.M. *Gravity: Newtonian, Post-Newtonian, Relativistic*; Cambridge University Press: Cambridge, UK, 2014, doi:10.1017/CBO9781139507486.
40. Kopeikin, S.M. Proper Motion of Binary Pulsars as a Source of Secular Variations of Orbital Parameters. *Astrophys. J. Lett.* **1996**, *467*, L93. [CrossRef]
41. Altschul, B. Limits on Neutron Lorentz Violation from Pulsar Timing. *Phys. Rev. D* **2007**, *75*, 023001. [CrossRef]
42. Tasson, J.D. Lorentz violation, gravitomagnetism, and intrinsic spin. *Phys. Rev. D* **2012**, *86*, 124021. [CrossRef]
43. Panjwani, H.; Carbone, L.; Speake, C.C. Laboratory Searches for Preferred Frame Effects: Ongoing Work and Results at Birmingham. In Proceedings of the 5th Meeting on CPT and Lorentz Symmetry (CPT 10), Bloomington, IN, USA, 28 June–2 July 2010; pp. 194–198.
44. Flowers, N.A.; Goodge, C.; Tasson, J.D. Superconducting-Gravimeter Tests of Local Lorentz Invariance. *Phys. Rev. Lett.* **2017**, *119*, 201101. [CrossRef] [PubMed]
45. Bourgoin, A.; Le Poncin-Lafitte, C.; Hees, A.; Bouquillon, S.; Francou, G.; Angonin, M.C. Lorentz Symmetry Violations from Matter-Gravity Couplings with Lunar Laser Ranging. *Phys. Rev. Lett.* **2017**, *119*, 201102. [CrossRef] [PubMed]

46. Schreck, M. Fermionic Lorentz violation and its implications for interferometric gravitational-wave detection. *Class. Quant. Grav.* **2017**, *34*, 135009. [CrossRef]
47. Kostelecký, V.A.; Tasson, J. Prospects for Large Relativity Violations in Matter-Gravity Couplings. *Phys. Rev. Lett.* **2009**, *102*, 010402. [CrossRef]
48. Hohensee, M.A.; Chu, S.; Peters, A.; Mueller, H. Equivalence Principle and Gravitational Redshift. *Phys. Rev. Lett.* **2011**, *106*, 151102. [CrossRef]
49. Hohensee, M.A.; Mueller, H.; Wiringa, R.B. Equivalence Principle and Bound Kinetic Energy. *Phys. Rev. Lett.* **2013**, *111*, 151102. [CrossRef]
50. Pihan-Le Bars, H.; Guerlin, C.; Lasseri, R.D.; Ebran, J.P.; Bailey, Q.G.; Bize, S.; Khan, E.; Wolf, P. Lorentz-symmetry test at Planck-scale suppression with nucleons in a spin-polarized ^{133}Cs cold atom clock. *Phys. Rev. D* **2017**, *95*, 075026. [CrossRef]
51. Damour, T.; Esposito-Farèse, G. Nonperturbative strong field effects in tensor-scalar theories of gravitation. *Phys. Rev. Lett.* **1993**, *70*, 2220–2223. [CrossRef] [PubMed]
52. Damour, T.; Esposito-Farèse, G. Tensor-scalar gravity and binary pulsar experiments. *Phys. Rev. D* **1996**, *54*, 1474–1491. [CrossRef] [PubMed]
53. Shao, L.; Sennett, N.; Buonanno, A.; Kramer, M.; Wex, N. Constraining nonperturbative strong-field effects in scalar-tensor gravity by combining pulsar timing and laser-interferometer gravitational-wave detectors. *Phys. Rev. X* **2017**, *7*, 041025. [CrossRef]
54. Zhao, J.; Shao, L.; Cao, Z.; Ma, B.Q. Reduced-order surrogate models for scalar-tensor gravity in the strong field and applications to binary pulsars and GW170817. *arXiv* **2019**, arXiv:1907.00780 .
55. Yunes, N.; Yagi, K.; Pretorius, F. Theoretical Physics Implications of the Binary Black-Hole Mergers GW150914 and GW151226. *Phys. Rev. D* **2016**, *94*, 084002. [CrossRef]
56. Sathyaprakash, B.S.; Buonanno, A.; Lehner, L.; Van Den Broeck, C.; Ajith, P.; Ghosh, A.; Chatziioannou, K.; Pani, P.; Puerrer, M.; Sotiriou, T.; et al. Extreme Gravity and Fundamental Physics. *arXiv* **2019**, arXiv:1903.09221.
57. Carson, Z.; Seymour, B.C.; Yagi, K. Future Prospects for Probing Scalar-Tensor Theories with Gravitational Waves from Mixed Binaries. *arXiv* **2019**, arXiv:1907.03897.
58. Kramer, M.; Backer, D.C.; Cordes, J.M.; Lazio, T.J.W.; Stappers, B.W.; Johnston, S. Strong-field tests of gravity using pulsars and black holes. *New Astron. Rev.* **2004**, *48*, 993–1002. [CrossRef]
59. Shao, L.; Stairs, I.H.; Antoniadis, J.; Deller, A.T.; Freire, P.C.C.; Hessels, J.W.T.; Janssen, G.H.; Kramer, M.; Kunz, J.; Lämmerzahl, C.; et al. Testing Gravity with Pulsars in the SKA Era. In Proceedings of the Advancing Astrophysics with the Square Kilometre Array, Giardini Naxos, Italy, 9–13 June 2014; Volume AASKA14, p. 042.
60. Bull, P.; Camera, S.; Kelley, K.; Padmanabhan, H.; Pritchard, J.; Raccanelli, A.; Riemer-Sørensen, S.; Shao, L.; Andrianomena, S.; Athanassoula, E.; et al. Fundamental Physics with the Square Kilometre Array. *arXiv* **2018**, arXiv:1810.02680.
61. Bailes, M.; Barr, E.; Bhat, N.D.R.; Brink, J.; Buchner, S.; Burgay, M.; Camilo, F.; Champion, D.J.; Hessels, J.; Janssen, G.H.; et al. MeerTime—The MeerKAT Key Science Program on Pulsar Timing. In Proceedings of the MeerKAT Science: On the Pathway to the SKA (MeerKAT2016), Stellenbosch, South Africa, 25–27 May 2016; Volume MeerKAT2016, p. 011. [CrossRef]

symmetry

MDPI

Article

Modifications to Plane Gravitational Waves from Minimal Lorentz Violation

Rui Xu

Kavli Institute for Astronomy and Astrophysics, Peking University, Beijing 100871, China; xuru@pku.edu.cn

Received: 11 October 2019; Accepted: 18 October 2019; Published: 21 October 2019

Abstract: General Relativity predicts two modes for plane gravitational waves. When a tiny violation of Lorentz invariance occurs, the two gravitational wave modes are modified. We use perturbation theory to study the detailed form of the modifications to the two gravitational wave modes from the minimal Lorentz-violation coupling. The perturbation solution for the metric fluctuation up to the first order in Lorentz violation is discussed. Then, we investigate the motions of test particles under the influence of the plane gravitational waves with Lorentz violation. First-order deviations from the usual motions are found.

Keywords: gravitational waves; Lorentz violation; standard-model extension; geodesic deviation

1. Introduction

General Relativity (GR), as the standard classical gravitational theory, has been making predictions consistent with all the terrestrial experiments and most of the astrophysical observations [1,2]. However, the fact that it is incompatible with quantum theory motivates ceaseless new tests and a large amount of alternative theories [3,4]. Lorentz invariance, being one of the fundamental principles in GR, has been suffering constant tests in various high-precision experiments and observations [5–9]. Especially, gravitational wave observations, providing us the unique access to strong-field environments, have recently put new stringent constraints on Lorentz violation based on the analysis of the modified dispersion relation of gravitational waves in the Standard-Model Extension (SME) framework [10,11].

The SME framework is a tool to study Lorentz violation in a model-independent way [12–18]. It incorporates all possible Lorentz-violation couplings into the Lagrangian density of GR and the Standard Model by employing the so called Lorentz-violation coefficients which can be measured or constrained with experimental data. The sector that describes gravity with Lorentz violation in vacuum is called the pure gravity sector of the SME [19,20], and it is the theoretical basis from which the modified dispersion relation of gravitational waves is derived [10,21].

Using the modified dispersion relation to constrain Lorentz violation marks the beginning of testing Lorentz invariance with gravitational wave observations [22]. As the number and sensitivity of gravitational wave observatories increase [23,24], we can extract more information about the incoming waves from the observed signals, including the polarization status of them. Recently, a detailed investigation on plane-wave solutions for arbitrary Lorentz violation in the pure gravitationa SME is carried out, and the modifications to the two polarization modes of the gravitational waves from coalescing compact binaries are considered [25]. Here we study a similar question but only with the simplest Lorentz-violation coupling in the pure gravity sector of the SME so that the calculations are more transparent. We have to point out that there are much bigger indicators of Lorentz violation [26] than what is described here. Therefore, our result is mostly pedagogical. In case it is to be used to constrain Lorentz violation in gravitational wave observations, a more comprehensive treatment to strain signals in gravitational-wave detectors is required.

We start with describing the basics of the minimal Lorentz-violation coupling [13,19] and show that a plane wave ansatz gives a naive modification to the usual plane wave solution in Section 2.

In Section 3, we generalize the naive modification to serve as a rigorous perturbation solution to the Lorentz-violation field equations. In Section 4, the perturbation solution is used to find the geodesic deviation of test particles on a ring under the effect of gravitational waves with Lorentz violation.

2. Plane Waves with Minimal Lorentz Violation

The Lorentz-violation couplings in the SME framework are constructed as coordinate scalars of the Lorentz-violation coefficients and conventional field operators. The simplest term in the pure gravity sector is [13,19]

$$\mathcal{L}^{(4)} = \frac{1}{16\pi G}(-uR + s^{\mu\nu}R^T_{\mu\nu} + t^{\alpha\beta\gamma\delta}C_{\alpha\beta\gamma\delta}), \tag{1}$$

where $u, s^{\mu\nu}$, and $t^{\alpha\beta\gamma\delta}$ are called the minimal Lorentz-violation coefficients as the coupling involves no derivatives of the Riemann tensor. $s^{\mu\nu}$ and $t^{\alpha\beta\gamma\delta}$ inherit the symetries and traceless property of the trace-free Ricci tensor, $R^T_{\mu\nu}$, and the Weyl conformal tensor, $C_{\alpha\beta\gamma\delta}$, separately. Note that the superscript 4 on $\mathcal{L}$ represents the mass dimension of the gravitational operators (including the gravitational constant factor G). Therefore, the minimal Lorentz-violation coefficients $u, s^{\mu\nu}$, and $t^{\alpha\beta\gamma\delta}$ are also called the Lorentz-violation coefficients with mass dimension $d = 4$.

Adding to the Einstein–Hilbert term, the Lagrangian density (1) gives modifications from minimal Lorentz-violation to the Einstein field equations. The details on linearizing the modified field equations and expressing them in terms of the background values of $u, s^{\mu\nu}$, and $t^{\alpha\beta\gamma\delta}$ are demonstrated in Ref. [19]. Here we just show the result which is the starting point of our calculation, namely the linearized vacuum field equations with minimal Lorentz-violation. They are

$$R_{\mu\nu} = \bar{s}^{\alpha\beta}R_{\alpha\mu\nu\beta}, \tag{2}$$

with $\bar{s}^{\alpha\beta}$ being the background value of $s^{\mu\nu}$. Note that the background values of u and $t^{\alpha\beta\gamma\delta}$ do not appear [19]. We also point out that the word "linearized" has two meanings here. One is the same as usual, namely the gravitational field is linearized. The second is that Equation (2) holds up to the first order in $\bar{s}^{\alpha\beta}$. There is no need to keep terms at higher orders in $\bar{s}^{\alpha\beta}$ because Lorentz violation should be tiny to be consistent with the experimental support for Lorentz invariance.

The dispersion relation implied by a generalized form of Equation (2) is studied in Ref. [27] to predict gravitational Čerenkov radiation from Lorentz violation. They proposed the modified harmonic gauge condition,

$$(\eta^{\lambda\kappa} + \bar{s}^{\lambda\kappa})\partial_\lambda h_{\kappa\mu} = \tfrac{1}{2}(\eta^{\lambda\kappa} + \bar{s}^{\lambda\kappa})\partial_\mu h_{\lambda\kappa}, \tag{3}$$

that simplifies the field equation (2) to

$$(\eta^{\alpha\beta} + \bar{s}^{\alpha\beta})\partial_\alpha\partial_\beta h_{\mu\nu} = 0, \tag{4}$$

where $h_{\mu\nu} = g_{\mu\nu} - \eta_{\mu\nu}$ is the fluctuation of the metric. Using the plane wave ansatz

$$h_{\mu\nu}(x) = A_{\mu\nu}e^{ikx}, \tag{5}$$

the modified dispersion relation up to the first order in $\bar{s}^{\alpha\beta}$ is found to be

$$k^0 = |\vec{k}| + \frac{1}{2}\frac{\bar{s}^{\alpha\beta}k_\alpha k_\beta}{|\vec{k}|}. \tag{6}$$

Namely the wave vector can be written as $k^\mu = (\omega + \delta\omega, \vec{k})$ with $\omega = |\vec{k}|$ and $\delta\omega = \frac{1}{2}\frac{\bar{s}^{\alpha\beta}k_\alpha k_\beta}{|\vec{k}|}$. Thus, the plane wave solution can be written as

$$h_{\mu\nu}(x) = A_{\mu\nu}e^{-i(\omega t - \vec{k}\cdot\vec{x})} - i\,\delta\omega\, t\, A_{\mu\nu}e^{-i(\omega t - \vec{k}\cdot\vec{x})} + \ldots. \tag{7}$$

The first term, $A_{\mu\nu}e^{-i(\omega t - \vec{k}\cdot x)}$, is apparently the plane wave solution in GR, and the rest consists of corrections from Lorentz violation. Up to the first order in $\bar{s}^{\alpha\beta}$, the correction is

$$h^{(1)}_{\mu\nu} = -i\,\delta\omega\, t\, A_{\mu\nu}e^{-i(\omega t - \vec{k}\cdot\vec{x})}. \tag{8}$$

As there is a factor of t in the amplitude of $h^{(1)}_{\mu\nu}$, the correction is only valid during a finite time period. The plane wave solution (7) is insufficient to describe the entire content of the Lorentz-violation modification to gravitational waves. However, Equation (8) provides us an insight into how the modification might look. In the next section, we will take the generalized form of Equation (8), which is

$$h^{(1)}_{\mu\nu} = C_{\mu\nu\alpha}x^\alpha e^{-i(\omega t - \vec{k}\cdot\vec{x})}, \tag{9}$$

as an ansatz to solve the field equation (4) up to the first order in $\bar{s}^{\alpha\beta}$. The constants $C_{\mu\nu\alpha}$ are going to be determined by the gauge condition (3) and the field equation (4). Note that the Lorentz-violation modification shown in Equation (9) applies only to a finite spacetime region as the coordinates x^α appear in the amplitude.

3. The Perturbation Solution

We seek a perturbation solution up to the first order in $\bar{s}^{\alpha\beta}$ for the field equation (4). To proceed, we assume that the zeroth-order plane wave travels along the z direction with the conventional wave vector

$$k^{(0)\mu} = (\omega, 0, 0, \omega), \tag{10}$$

and that its amplitude $A_{\mu\nu}$ takes the usual form

$$A_{\mu\nu} = \begin{pmatrix} 0 & 0 & 0 & 0 \\ 0 & A_{11} & A_{12} & 0 \\ 0 & A_{12} & -A_{11} & 0 \\ 0 & 0 & 0 & 0 \end{pmatrix}, \tag{11}$$

where A_{11} is the amplitude of the plus wave and A_{12} is the amplitude of the cross wave. By substituting

$$h_{\mu\nu}(x) = A_{\mu\nu}e^{-i(\omega t - kz)} + C_{\mu\nu\alpha}x^\alpha e^{-i(\omega t - kz)}, \tag{12}$$

into the field quation (4) and keeping only the first-order terms, we have

$$2C_{\mu\nu\alpha}ik^{(0)\alpha} = \bar{s}^{\alpha\beta}k^{(0)}_\alpha k^{(0)}_\beta A_{\mu\nu}. \tag{13}$$

Writing the above equations explicitly, they are

$$C_{\mu\nu 0} + C_{\mu\nu 3} = -\frac{i\omega}{2}(\bar{s}^{00} - 2\bar{s}^{03} + \bar{s}^{33})A_{\mu\nu}. \tag{14}$$

In addition, using Equation (12) in the gauge condition (3), up to the first order we have

$$\eta^{\lambda\kappa}C_{\kappa\mu\alpha}ik^{(0)}_\lambda x^\alpha + \eta^{\lambda\kappa}C_{\kappa\mu\lambda} + \bar{s}^{\lambda\kappa}ik^{(0)}_\lambda A_{\kappa\mu} = \frac{1}{2}\left(C_\alpha ik^{(0)}_\mu x^\alpha + C_\mu + \bar{s}^{\lambda\kappa}ik^{(0)}_\mu A_{\lambda\kappa}\right), \tag{15}$$

where $C_\alpha = \eta^{\mu\nu} C_{\mu\nu\alpha}$. The relations (15) imply two sets of equations:

$$\eta^{\lambda\kappa} C_{\kappa\mu\alpha} k_\lambda^{(0)} - \tfrac{1}{2} C_\alpha k_\mu^{(0)} = 0, \tag{16}$$

and

$$\eta^{\lambda\kappa} C_{\kappa\mu\lambda} - \tfrac{1}{2} C_\mu = \tfrac{i}{2} \bar{s}^{\lambda\kappa} k_\mu^{(0)} A_{\lambda\kappa} - i \bar{s}^{\lambda\kappa} k_\lambda^{(0)} A_{\kappa\mu}. \tag{17}$$

Using the expressions (10) and (11), we find that Equation (16) can be simplified to

$$\begin{aligned} C_{00\alpha} + 2C_{03\alpha} + C_{33\alpha} &= 0, \\ C_{11\alpha} + C_{22\alpha} &= 0, \\ C_{01\alpha} + C_{31\alpha} &= 0, \\ C_{02\alpha} + C_{32\alpha} &= 0, \end{aligned} \tag{18}$$

and that Equation (17) can be simplified to

$$\begin{aligned} C_{011} + C_{022} &= -\tfrac{1}{2} i\omega \left((\bar{s}^{11} - \bar{s}^{22}) A_{11} + 2\bar{s}^{12} A_{12} \right), \\ C_{111} + C_{122} + \tfrac{1}{2}(C_{001} - C_{331}) &= i\omega A_{11} (\bar{s}^{01} - \bar{s}^{31}) + i\omega A_{12} (\bar{s}^{02} - \bar{s}^{32}), \\ C_{121} - C_{112} + \tfrac{1}{2}(C_{002} - C_{332}) &= i\omega A_{12} (\bar{s}^{01} - \bar{s}^{31}) - i\omega A_{11} (\bar{s}^{02} - \bar{s}^{32}). \end{aligned} \tag{19}$$

Note that Equation (17) turns out to have only 3 independent equations.

Equation (19) shows that there are 6 independent components in the first-order solution $h_{\mu\nu}^{(1)}$, which can be written as

$$h_{\mu\nu}^{(1)} = \begin{pmatrix} h_{00}^{(1)} & h_{01}^{(1)} & h_{02}^{(1)} & -\tfrac{1}{2}(h_{00}^{(1)} + h_{33}^{(1)}) \\ h_{01}^{(1)} & h_{11}^{(1)} & h_{12}^{(1)} & -h_{01}^{(1)} \\ h_{02}^{(1)} & h_{12}^{(1)} & -h_{11}^{(1)} & -h_{02}^{(1)} \\ -\tfrac{1}{2}(h_{00}^{(1)} + h_{33}^{(1)}) & -h_{01}^{(1)} & -h_{02}^{(1)} & h_{33}^{(1)} \end{pmatrix}. \tag{20}$$

The 6 independent components are easily divided into 3 groups, $\{h_{11}^{(1)}, h_{12}^{(1)}\}$, $\{h_{00}^{(1)}, h_{33}^{(1)}\}$, and $\{h_{01}^{(1)}, h_{02}^{(1)}\}$. The remaining equations in (14), and (20) are insufficient to determine any of them. This indicates that the ansatz (9) does not lead to a unique first-order solution. We need extra information to fix $h_{\mu\nu}^{(1)}$. Next, we discuss the solutions for $\{h_{11}^{(1)}, h_{12}^{(1)}\}$, $\{h_{00}^{(1)}, h_{33}^{(1)}\}$ and $\{h_{01}^{(1)}, h_{02}^{(1)}\}$ separately.

3.1. $\{h_{11}^{(1)}, h_{12}^{(1)}\}$

We expect these two components recover the correction (8). This is indeed the case if we take all the components of $C_{11\alpha}$ and $C_{12\alpha}$ to be zero except for

$$\begin{aligned} C_{110} &= -\tfrac{i\omega}{2} (\bar{s}^{00} - 2\bar{s}^{03} + \bar{s}^{33}) A_{11}, \\ C_{120} &= -\tfrac{i\omega}{2} (\bar{s}^{00} - 2\bar{s}^{03} + \bar{s}^{33}) A_{12}. \end{aligned} \tag{21}$$

In this way, $h_{11}^{(1)}$ and $h_{12}^{(1)}$ are fixed, and the dispersion relation (6) can be recovered in the perturbation solution.

3.2. $\{h^{(1)}_{00}, h^{(1)}_{33}\}$

With $C_{111} = C_{112} = C_{121} = C_{122} = 0$, we have

$$\begin{aligned} C_{001} - C_{331} &= 2i\omega A_{11}(\bar{s}^{01} - \bar{s}^{31}) + 2i\omega A_{12}(\bar{s}^{02} - \bar{s}^{32}), \\ C_{002} - C_{332} &= 2i\omega A_{12}(\bar{s}^{01} - \bar{s}^{31}) - 2i\omega A_{11}(\bar{s}^{02} - \bar{s}^{32}). \end{aligned} \tag{22}$$

It turns out the combinations $C_{001} - C_{331}$ and $C_{002} - C_{332}$ are the only terms involving $C_{00\alpha}$ and $C_{33\alpha}$ in the first-order Riemann tensor (see the Appendix A). Therefore, without any ambiguity in observables, we can safely assume all the components of $C_{00\alpha}$ and $C_{33\alpha}$ vanishing except for C_{001} and C_{002}, which are given by Equation (22).

3.3. $\{h^{(1)}_{01}, h^{(1)}_{02}\}$

In the Appendix A, we can see that C_{010}, C_{013}, C_{020}, and C_{023} do not appear in the first-order Riemann tensor. Therefore, they can be taken as zero. However, C_{011} and C_{022} appear, and they do not appear as the combination $C_{011} + C_{022}$ as shown in Equation (20). In addition, C_{012} and C_{021} also show up in the first-order Riemann tensor. Namely, we have one equation in (20) to use but 4 unknowns, C_{011}, C_{012}, C_{021}, and C_{022}, to determine. The inadequacy is likely from the fact that we are missing certain information about the specific dynamic model of the Lorentz-violation coefficient $s^{\alpha\beta}$. In other words, we expect $s^{\alpha\beta}$ to have its own field equations with the metric involved. Then, when $s^{\alpha\beta}$ is approximated by its background value $\bar{s}^{\alpha\beta}$, some of these field equations degenerate to constraints on the metric though most of them vanish trivially.

Building a specific dynamic model for $s^{\alpha\beta}$ simply lies beyond the scope of the present work. For the calculation in the next section, we decide to choose the simplest solution for $h^{(1)}_{01}$ and $h^{(1)}_{02}$, by which we mean that all the components of $C_{01\alpha}$ and $C_{02\alpha}$ vanish except for

$$C_{011} = -\tfrac{1}{2}i\omega\big((\bar{s}^{11} - \bar{s}^{22})A_{11} + 2\bar{s}^{12}A_{12}\big). \tag{23}$$

4. Geodesic Deviation

Now we use the above first-order solution to calculate the effects of Lorentz violation on the motions of test particles when plane gravitational waves pass through. Similarly to the usual case, it is illustrative to consider a ring of test particles whose initial positions form a circle

$$(X(0))^2 + (Y(0))^2 = d^2, \tag{24}$$

in a local inertial frame with local coordinates $\{X, Y, Z\}$. Assuming the local coordinates are aligned with the general coordinates $\{x, y, z\}$, then the nonrelativistic geodesic deviation equations that determine the motions of the test particles in the local frame are [28]

$$\begin{aligned} \tfrac{d^2X}{dt^2} &= -R_{0101}X(0) - R_{0102}Y(0) - R_{0103}Z(0), \\ \tfrac{d^2Y}{dt^2} &= -R_{0201}X(0) - R_{0202}Y(0) - R_{0203}Z(0), \\ \tfrac{d^2Z}{dt^2} &= -R_{0301}X(0) - R_{0302}Y(0) - R_{0303}Z(0). \end{aligned} \tag{25}$$

The zeroth-order solution for $X(t)$, $Y(t)$, and $Z(t)$ is the usual deformation

$$\begin{aligned} X^{(0)}(t) - X(0) &= \tfrac{1}{2}\big(A_{11}X(0) + A_{12}Y(0)\big)(e^{-i\omega t} - 1), \\ Y^{(0)}(t) - Y(0) &= \tfrac{1}{2}\big(A_{12}X(0) - A_{11}Y(0)\big)(e^{-i\omega t} - 1), \\ Z^{(0)}(t) - Z(0) &= 0. \end{aligned} \tag{26}$$

Note that we have assumed that the local frame is moving along the geodesic $x(t) = y(t) = z(t) = 0$. The first-order solution turns out to be

$$\begin{aligned} X^{(1)}(t) &= -\tfrac{1}{\omega^2}(\alpha_X + \beta_X(\omega t - 2i))e^{-i\omega t} + \tfrac{1}{\omega^2}(\alpha_X - 2i\beta_X), \\ Y^{(1)}(t) &= -\tfrac{1}{\omega^2}(\alpha_Y + \beta_Y(\omega t - 2i))e^{-i\omega t} + \tfrac{1}{\omega^2}(\alpha_Y - 2i\beta_Y), \\ Z^{(1)}(t) &= -\tfrac{1}{\omega^2}\alpha_Z e^{-i\omega t} + \tfrac{1}{\omega^2}\alpha_Z, \end{aligned} \tag{27}$$

where

$$\begin{aligned} \alpha_X &= -i\omega(C_{110} - C_{011})X(0) - i\omega C_{120}Y(0) + \tfrac{1}{4}i\omega C_{001}Z(0), \\ \beta_X &= -\tfrac{1}{2}\omega C_{110}X(0) - \tfrac{1}{2}\omega C_{120}Y(0), \\ \alpha_Y &= -i\omega C_{120}X(0) + i\omega C_{110}Y(0) + \tfrac{1}{4}i\omega C_{002}Z(0), \\ \beta_Y &= -\tfrac{1}{2}\omega C_{120}X(0) + \tfrac{1}{2}\omega C_{110}Y(0), \\ \alpha_Z &= \tfrac{1}{4}i\omega C_{001}X(0) + \tfrac{1}{4}i\omega C_{002}Y(0). \end{aligned} \tag{28}$$

The solution (27) as well as Equation (28) is written with the understanding that only the real parts are taken.

The most notable correction is that Lorentz violation causes an oscillation along the z direction in general, which does not happen in the case of the usual plane gravitational waves. Then, for the ring of the test particles in the XY plane, we find that the shape is still deformed into ellipses. But the semi axes are corrected by Lorentz violation. Specifically speaking, when A_{11} is real and $A_{12} = 0$, the semi axes of the ellipse at time t are

$$\begin{aligned} a &= d(1 + \tfrac{1}{2}A_{11}(\cos\omega t - 1) - \tfrac{1}{2}A_{11}(\bar{s}^{11} - \bar{s}^{22})(\cos\omega t - 1) - \tfrac{1}{4}A_{11}(\bar{s}^{00} - 2\bar{s}^{03} + \bar{s}^{33})\omega t \sin\omega t), \\ b &= d(1 - \tfrac{1}{2}A_{11}(\cos\omega t - 1) + \tfrac{1}{4}A_{11}(\bar{s}^{00} - 2\bar{s}^{03} + \bar{s}^{33})\omega t \sin\omega t); \end{aligned} \tag{29}$$

when A_{12} is real and $A_{11} = 0$, the semi axes of the ellipse at time t are

$$\begin{aligned} a &= d(1 + A_{12}(\cos\omega t - 1) - A_{12}\bar{s}^{12}(\cos\omega t - 1) - \tfrac{1}{2}A_{12}(\bar{s}^{00} - 2\bar{s}^{03} + \bar{s}^{33})\omega t \sin\omega t), \\ b &= d(1 - A_{12}(\cos\omega t - 1) - A_{12}\bar{s}^{12}(\cos\omega t - 1) + \tfrac{1}{2}A_{12}(\bar{s}^{00} - 2\bar{s}^{03} + \bar{s}^{33})\omega t \sin\omega t). \end{aligned} \tag{30}$$

Last but not least, we point out that when A_{12} is real and $A_{11} = 0$, the rotation angle of the ellipses from the standard position

$$\tfrac{X^2}{a^2} + \tfrac{Y^2}{b^2} = 1, \tag{31}$$

is not $\pm\frac{\pi}{4}$ any more. A time-independent deviation of $\frac{1}{2}\bar{s}^{12}$ occurs in the presence of Lorentz violation.

5. Conclusions

We used the ansatz (9) to find the correction to plane gravitational waves from minimal Lorentz violation. It was shown that up to the first order in Lorentz violation, the correction, $h^{(1)}_{\mu\nu}$, has 6 independent components, with 4 of them fixed in the SME framework. To determine the remaining two components, extra information about the dynamics of the Lorentz-violation coefficient $s^{\alpha\beta}$ is necessary. This requires treating $s^{\alpha\beta}$ as a dynamic field and assigning it a kinetic term in the Lagrangian density. This lies beyond the scope of the present work.

Then, to demonstrate the effects of Lorentz violation on the motions of test particles under the influence of plane gravitational waves, we artificially fixed the two undetermined components of $h^{(1)}_{\mu\nu}$. Together with the other 4 determined components, two notable effects were found. One is the oscillation of a test particle along the propagating direction of the gravitational waves, and the other is the deviation from $\pm\frac{\pi}{4}$ for the rotation angle of the deformed ellipse in the presence of the cross wave. Note that the amplitude of the oscillation along the Z-direction is proportional to the amplitude of the zeroth-order gravitational wave but suppressed by the components of the Lorentz-violation coefficient $\bar{s}^{\alpha\beta}$. Taking the current best bound of 10^{-15} [11] on $\bar{s}^{\alpha\beta}$ into consideration, it is unlikely that this oscillation provides a viable test of Lorentz violation even in the near future. On the other hand, as we are getting access to the polarization information of incoming gravitational waves with more and more detectors in construction, the deviation of the rotation angle suggests a Lorentz-violation phase difference between the two polarization modes to test in future observations of polarized gravitational waves. To conduct such tests, a more comprehensive treatment in the context of existing and future gravitational-wave detectors is required, which deserves another paper for investigation.

Funding: This research received no external funding.

Acknowledgments: R.X. is thankful to Alan Kostelecký, Lijing Shao, and Jay Tasson for valuable comments.

Conflicts of Interest: The author declares no conflict of interest.

Appendix A. The First-Order Riemann Tensor

The first-order Riemann tensor is calculated by

$$R^{(1)}_{\alpha\beta\gamma\delta} = \tfrac{1}{2}(\partial_\gamma\partial_\beta h^{(1)}_{\alpha\delta} + \partial_\alpha\partial_\delta h^{(1)}_{\beta\gamma} - \partial_\gamma\partial_\alpha h^{(1)}_{\beta\delta} - \partial_\delta\partial_\beta h^{(1)}_{\alpha\gamma}). \tag{A1}$$

Plugging Equation (9) into it, and using $\vec{k} = (0, 0, \omega)$, we find

$$\begin{aligned}
R^{(1)}_{0101} &= \tfrac{1}{2}(2i\omega(C_{110} - C_{011}) + \omega^2 C_{11\alpha}x^\alpha)e^{-i(\omega t - \vec{k}\cdot\vec{x})}, \\
R^{(1)}_{0102} &= \tfrac{1}{2}(i\omega(2C_{120} - C_{021} - C_{012}) + \omega^2 C_{12\alpha}x^\alpha)e^{-i(\omega t - \vec{k}\cdot\vec{x})}, \\
R^{(1)}_{0103} &= -\tfrac{1}{4}i\omega(C_{001} - C_{331})e^{-i(\omega t - \vec{k}\cdot\vec{x})}, \\
R^{(1)}_{0202} &= -\tfrac{1}{2}(2i\omega(C_{110} + C_{022}) + \omega^2 C_{11\alpha}x^\alpha)e^{-i(\omega t - \vec{k}\cdot\vec{x})}, \\
R^{(1)}_{0203} &= -\tfrac{1}{4}i\omega(C_{002} - C_{332})e^{-i(\omega t - \vec{k}\cdot\vec{x})}, \\
R^{(1)}_{0303} &= 0,
\end{aligned} \tag{A2}$$

$$\begin{aligned}
R^{(1)}_{0112} &= -\tfrac{1}{2}i\omega(C_{112} - C_{121})e^{-i(\omega t - \vec{k}\cdot\vec{x})}, \\
R^{(1)}_{0113} &= \tfrac{1}{2}(i\omega(C_{110} - C_{113} - 2C_{011}) + \omega^2 C_{11\alpha}x^\alpha)e^{-i(\omega t - \vec{k}\cdot\vec{x})}, \\
R^{(1)}_{0123} &= -\tfrac{1}{2}i\omega C_{012}e^{-i(\omega t - \vec{k}\cdot\vec{x})}, \\
R^{(1)}_{0212} &= -\tfrac{1}{2}i\omega(C_{111} + C_{122})e^{-i(\omega t - \vec{k}\cdot\vec{x})}, \\
R^{(1)}_{0213} &= \tfrac{1}{2}(i\omega(C_{120} - C_{123} - C_{021} - C_{012}) + \omega^2 C_{12\alpha}x^\alpha)e^{-i(\omega t - \vec{k}\cdot\vec{x})}, \\
R^{(1)}_{0223} &= \tfrac{1}{2}(-i\omega(C_{110} - C_{113} + 2C_{022}) - \omega^2 C_{11\alpha}x^\alpha)e^{-i(\omega t - \vec{k}\cdot\vec{x})}, \\
R^{(1)}_{0313} &= -\tfrac{1}{4}i\omega(C_{001} - C_{331})e^{-i(\omega t - \vec{k}\cdot\vec{x})}, \\
R^{(1)}_{0323} &= -\tfrac{1}{4}i\omega(C_{002} - C_{332})e^{-i(\omega t - \vec{k}\cdot\vec{x})},
\end{aligned} \tag{A3}$$

and

$$\begin{aligned}
R^{(1)}_{1212} &= 0, \\
R^{(1)}_{1213} &= -\tfrac{1}{2}i\omega(C_{112} - C_{121})e^{-i(\omega t - \vec{k}\cdot\vec{x})}, \\
R^{(1)}_{1223} &= -\tfrac{1}{2}i\omega(C_{111} + C_{122})e^{-i(\omega t - \vec{k}\cdot\vec{x})}, \\
R^{(1)}_{1313} &= \tfrac{1}{2}\big(-2i\omega(C_{113} + C_{011}) + \omega^2 C_{11\alpha}x^{\alpha}\big)e^{-i(\omega t - \vec{k}\cdot\vec{x})}, \\
R^{(1)}_{1323} &= \tfrac{1}{2}\big(-i\omega(C_{012} + C_{021} + 2C_{123}) + \omega^2 C_{12\alpha}x^{\alpha}\big)e^{-i(\omega t - \vec{k}\cdot\vec{x})}, \\
R^{(1)}_{2323} &= \tfrac{1}{2}\big(2i\omega(C_{113} - C_{022}) - \omega^2 C_{11\alpha}x^{\alpha}\big)e^{-i(\omega t - \vec{k}\cdot\vec{x})}.
\end{aligned} \tag{A4}$$

References

1. Debono, I.; Smoot, G.F. General Relativity and Cosmology: Unsolved Questions and Future Directions. *Universe* **2016**, *2*, 23. [CrossRef]
2. Vishwakarma, R.G. Einstein and Beyond: A Critical Perspective on General Relativity. *Universe* **2016**, *2*, 11. [CrossRef]
3. Will, C.M. The confrontation between general relativity and experiment. *Living Rev. Relativ.* **2014**, *14*, 4. [CrossRef] [PubMed]
4. Berti, E.; Barausse, E.; Cardoso, V.; Gualtieri, L.; Pani, P.; Sperhake, U.; Stein, L.C.; Wex, N.; Yagi, K.; Baker, T.; et al. Testing General Relativity with Present and Future Astrophysical Observations. *Class. Quantum Grav.* **2015**, *32*, 243001. [CrossRef]
5. Kostelecký, V.A.; Russell, N. Data tables for Lorentz and CPT violation. *Rev. Mod. Phys.* **2011**, *83*, 11. [CrossRef]
6. Battat, J.B.R.; Chandler, J.F.; Stubbs, C.W. Testing for Lorentz Violation: Constraints on Standard-Model Extension Parameters via Lunar Laser Ranging. *Phys. Rev. Lett.* **2007**, *99*, 241103. [CrossRef]
7. Muller, H.; Chiow, S.-W.; Herrmann, S.; Chu, S.; Chung, K.-Y. Atom Interferometry tests of the isotropy of post-Newtonian gravity. *Phys. Rev. Lett.* **2008**, *100*, 031101. [CrossRef]
8. Shao, L. Tests of local Lorentz invariance violation of gravity in the standard model extension with pulsars. *Phys. Rev. Lett.* **2014**, *112*, 111103. [CrossRef]
9. Shao, C.-G.; Tan, Y.-J.; Tan, W.-H.; Yang, S.-Q.; Luo, J.; Tobar, M.E.; Bailey, Q.G.; Long, J.C.; Weisman, E.; Xu, R.; et al. Combined search for Lorentz violation in short-range gravity. *Phys. Rev. Lett.* **2016**, *117*, 071102. [CrossRef]
10. Kostelecký, V.A.; Mewes, M. Testing local Lorentz invariance with gravitational waves. *Phys. Lett. B* **2016**, *757*, 510. [CrossRef]
11. Abbott, B.P.; Abbott, R.; Abbott, T.D.; Acernese, R.; Ackley, K.; Adams, C.; Adams, T.; Addesso, P.; Adhikari, R.X.; Adya, V.B.; et al. Gravitational waves and gamma-rays from a binary Neutron star merger: GW170817 and GRB 170817A. *Astrophys. J. Lett.* **2017**, *848*, L13. [CrossRef]
12. Colladay, D.; Kostelecký, V.A. Lorentz-violating extension of the Standard Model. *Phys. Rev. D* **1998**, *58*, 116002. [CrossRef]
13. Kostelecký, V.A. Gravity, Lorentz violation, and the Standard Model. *Phys. Rev. D* **2004**, *69*, 105009. [CrossRef]
14. Kostelecký, V.A.; Mewes, M. Electrodynamics with Lorentz-violating operators of arbitrary dimension. *Phys. Rev. D* **2009**, *80*, 015020. [CrossRef]
15. Kostelecký, V.A.; Tasson, J.D. Matter-gravity couplings and Lorentz violation. *Phys. Rev. D* **2011**, *83*, 016013. [CrossRef]
16. Kostelecký, V.A.; Mewes, M. Neutrinos with Lorentz-violating operators of arbitrary dimension. *Phys. Rev. D* **2012**, *85*, 096005. [CrossRef]
17. Kostelecký, V.A.; Mewes, M. Fermions with Lorentz-violating operators of arbitrary dimension. *Phys. Rev. D* **2013**, *88*, 096006. [CrossRef]
18. Tasson, J.D. What do we know about Lorentz invariance? *Rep. Prog. Phys.* **2014**, *77*, 062901. [CrossRef]
19. Bailey, Q.G.; Kostelecký, V.A. Signals for Lorentz violation in post-newtonian gravity. *Phys. Rev. D* **2006**, *74*, 045001. [CrossRef]

20. Bailey, Q.G.; Kostelecký, V.A.; Xu, R. Short-range gravity and Lorentz violation. *Phys. Rev. D* **2015**, *91*, 022006. [CrossRef]
21. Kostelecký, V.A.; Mewes, M. Lorentz and Diffeomorphism Violations in Linearized Gravity. *Phys. Lett. B* **2018**, *779*, 136. [CrossRef]
22. Cervantes-Cota, J.L.; Galindo-Uribarri, S.; Smoot, G.F. A Brief History of Gravitational Waves. *Universe* **2016**, *2*, 22. [CrossRef]
23. Akutsu, T.; Ando, M.; Arai, K.; Arai, Y.; Araki, S.; Araya, A.; Aritomi, N.; Asada, H.; Aso, Y.; Atsuta, S.; et al. KAGRA: 2.5 Generation Interferometric Gravitational Wave Detector. *arXiv* **2018**, arXiv:1811.08079.
24. Abbott, B.P.; Abbott, R.; Abbott, T.D.; Abernathy, M.R.; Acernese, F.; Ackley, K.; Adams, C.; Adams, T.; Addesso, P.; Adhikari, R.X.; et al. LIGO scientific collaboration and virgo collaboration. Gravitational wave astronomy with LIGO and similar detectors in the next decade. *arXiv* **2019**, arXiv:1904.03187.
25. Mewes, M. Signals for Lorentz violation in gravitational waves. *Phys. Rev. D* **2019**, *99*, 104062. [CrossRef]
26. Sotiriou, T.P. Detecting Lorentz Violations with Gravitational Waves from Black Hole Binaries. *Phys. Rev. Lett.* **2018**, *120*, 041104. [CrossRef]
27. Kostelecký, V.A.; Tasson, J.D. Constraints on Lorentz violation from gravitational Čerenkov radiation. *Phys. Lett. B* **2015**, *749*, 551. [CrossRef]
28. Poisson E.; Will, C.M. *Gravity*; Cambridge University Press: Cambridge, UK, 2014.

symmetry

MDPI

Brief Report

Symmetric Criticality and Magnetic Monopoles in General Relativity

Joel Franklin

Physics Department, Reed College, Portland, OR 97202, USA; jfrankli@reed.edu

Received: 6 June 2019; Accepted: 26 June 2019; Published: 1 June 2019

Abstract: The Weyl method for finding solutions in general relativity using symmetry by varying an action with respect to a reduced set of field variables is known to fail in some cases. We add to the list of failures by considering an application of the Weyl method to a magnetically charged spherically symmetric source, obtaining an incorrect geometry. This is surprising, because the same method, applied to electrically charged central bodies correctly produces the Reissner-Nordström spacetime.

Keywords: Weyl method; Palais principle of symmetric criticality; solutions to Einstein's equations; magnetic monopole

1. Introduction

We often use symmetry to simplify the field equations of general relativity (GR) and help solve them. There is a particular approach, the Weyl method [1], that benefits from an early application of assumed symmetry and can lead to striking simplification. The method has been used to successfully generate the spherically symmetric vacuum spacetime of general relativity, its first application. It has also been applied to modified theories like GR with cosmological constant, Einstein-Gauss-Bonnet gravity [2] and conformal gravity, all of which are developed and/or reviewed in Reference [3]. For axial symmetry, the $2+1$ dimensional "Kerr" solution for gravity with (negative) cosmological constant (BTZ) is obtained in Reference [3], with $3+1$ dimensional Kerr obtained using a targeted form of the technique in Reference [4].

But we must be careful, the method does not always work, as was detailed in Reference [5]. In this note, we review the method, providing some of its successful examples and discuss its failure in specific cases. We show that while the method is successful in finding the spacetime associated with an electrically charged spherical mass, it fails when the electric charge is replaced by magnetic charge (i.e., a magnetic monopole).

2. The Weyl Method

The Weyl method refers to the approach, invented and advertised by Weyl in Reference [1], of using information, in particular symmetry information, prior to varying an action in order to reduce the number, and simplify the form, of the field equations. Spherical symmetry in the Einstein-Hilbert action provides a good first example. Starting from the spherically symmetric line element, with two unknown function of r, the radial coordinate,

$$ds^2 = -A(r)dt^2 + B(r)dr^2 + r^2\left(d\theta^2 + \sin^2\theta d\phi^2\right), \quad (1)$$

we can form the Lagrangian for the action (primes indicate r-derivatives),

$$\mathcal{L} = \sqrt{-g}R = \frac{\sin\theta}{2\left(A(r)B(r)\right)^{3/2}}\left[r^2B(r)A'(r)^2 + 4A(r)^2\left(-B(r) + B(r)^2 + rB'(r)\right)\right.$$
$$\left. + rA(r)\left(rA'(r)B'(r) - 2B(r)\left(2A'(r) + rA''(r)\right)\right)\right], \tag{2}$$

and then we can use the Euler-Lagrange equations for $A(r)$ and $B(r)$,

$$0 = \frac{\partial\mathcal{L}}{\partial A(r)} - \frac{d}{dr}\left(\frac{\partial\mathcal{L}}{\partial A'(r)}\right) + \frac{d^2}{dr^2}\left(\frac{\partial\mathcal{L}}{\partial A''(r)}\right) = \frac{\sin\theta}{\sqrt{A(r)B(r)^3}}\left[B(r)\left(-1 + B(r)\right) + rB'(r)\right]$$
$$0 = \frac{\partial\mathcal{L}}{\partial B(r)} - \frac{d}{dr}\left(\frac{\partial\mathcal{L}}{\partial B'(r)}\right) = \frac{\sin\theta}{\sqrt{A(r)B(r)^3}}\left[A(r)\left(-1 + B(r)\right) - rA'(r)\right] \tag{3}$$

to find $A(r)$ and $B(r)$. Solving the top equation for $B(r)$, we get

$$B(r) = \frac{1}{1 - \frac{\alpha}{r}} \tag{4}$$

for constant α. Then using this in the second equation, we can solve for $A(r)$,

$$A(r) = 1 - \frac{\alpha}{r}. \tag{5}$$

We have recovered the Schwarzschild solution, with constant α awaiting its usual physical interpretation, $\alpha = 2M$, with $G \to 1$, $c \to 1$.

The beauty of the Weyl approach is that the assumed form of the line element can simplify (or complexify) the field equations for the unknown functions. For example, if we started with the two-function ($a(r)$, $b(r)$ now) line element as in Reference [3],

$$ds^2 = -a(r)b(r)^2dt^2 + 1/a(r)dr^2 + r^2\left(d\theta^2 + \sin^2\theta d\phi^2\right) \tag{6}$$

motivated by, for example, the single-function form of the determinant $\sqrt{-g} = b(r)r^2\sin\theta$ piece of the action, then the Lagrangian is

$$\mathcal{L} = -\sin\theta\left[b(r)\left(-2r + \left(r^2a(r)\right)'\right)' + \left(2r^2a(r)b'(r)\right)' + r^2a'(r)b'(r)\right] \tag{7}$$

with field equations (obtainable even by dropping the total r-derivative in $\mathcal{L}$),

$$2r\sin\theta b'(r) = 0 \qquad 2\sin\theta\left(1 - (ra(r))'\right) = 0 \tag{8}$$

a decoupled set that's even easier to solve than those in (3) and leads, of course, to the same Schwarzschild spacetime.

3. Symmetric Criticality

There is a problem with the Weyl approach, one that goes back to the idea of action variation itself. Symmetries can be applied at the level of a field equation and lead to correct simplifications. Indeed, the symmetry of a solution is implied by the form of the field equation and (more importantly), the boundary conditions we impose on its solutions. Simplifications of this sort belong to the PDE problem that the field equations and boundaries define. But any information that derives from the field equations must be treated carefully when used prior to varying an action, that is, prior to developing the field equations, precisely what Weyl invites us to do.

As a reductio ad absurdum example from classical mechanics, suppose we take the free particle action in one dimension,

$$S[x(t)] = \int_{t_0}^{t_f} \frac{1}{2} m \dot{x}(t)^2 dt \tag{9}$$

and vary the action to get the equation of motion, $m\ddot{x}(t) = 0$ from which we learn that $x(t) = ft + g$ for constants f and g. If we insert this solution back into the action, we get

$$S = \int_{t_0}^{t_f} \frac{1}{2} m f^2 dt \tag{10}$$

which cannot itself be varied to recover a valid equation of motion governing $x(t)$. This is the logic that shows the potential flaw in the Weyl procedure. We have fixed all the degrees of freedom by solving the equation of motion, leaving us with nothing to vary in the action when that solution has been introduced.

The previous example is contrived and extreme but consider the slightly more disguised error in the following: We note that for the Schwarzschild solution (4) and (5), $B(r) = 1/A(r)$ (this is what suggests the two-function form of the line element in (6)). Suppose we use that information in developing the Lagrangian, that is, start with the line element

$$ds^2 = -a(r)dt^2 + 1/a(r)dr^2 + r^2\left(d\theta^2 + \sin^2\theta d\phi^2\right). \tag{11}$$

Then the Lagrangian becomes

$$\mathcal{L} = -\sin\theta\left(-2 + 2a(r) + 4ra'(r) + r^2 a''(r)\right) = -\frac{d}{dr}\left[\sin\theta\left(-2r + \left(r^2 a(r)\right)'\right)\right] \tag{12}$$

which, since it is a total derivative, leads to a trivial field equation ($0 = 0$) leaving $a(r)$ unconstrained. One might naïvely conclude that any function $a(r)$ solves the field equation in the spherically symmetric case. This is, of course, incorrect. The actual field equation, Einstein's in vacuum, $R_{\mu\nu} = 0$, has non-zero entries:

$$R_{00} = \frac{a(r)}{2}\left(\frac{2a'(r)}{r} + a''(r)\right) \qquad R_{rr} = -R_{00}/a(r)^2 \qquad R_{\theta\theta} = 1 - (ra(r))' \qquad R_{\phi\phi} = \sin^2\theta R_{\theta\theta}, \tag{13}$$

and these are solved by the usual $a(r) = 1 - \alpha/r$. While we can start with (11) and get the correct result from the field equations themselves, we have used *too much* simplifying information to recover that result from the Weyl method [6]. It is easy to go back and check that a solution obtained via the Weyl method is valid by running it through Einstein's equation. What is more difficult is to determine, a priori, whether a particular simplifying assumption will lead to problems. The equivalence of "varying an action, then imposing symmetry assumptions" and "imposing symmetry assumptions and then varying an action" is an example of Palais' "principle of symmetric criticality" [5]. He cautions that the principle is not universal and the current case provides an example of its failure.

Another case in which the principle fails is in establishing Birkhoff's theorem in general relativity. Birkhoff's theorem says that the spherically symmetric vacuum solution to Einstein's equation (Schwarzschild) is static, with no time dependence. If you started with an ansatz like (1) but allowed the functions A and B to depend on time, you would find no constraint on their temporal dependence using the Weyl approach, while Einstein's equations explicitly require $\dot{A}(r,t) = 0 = \dot{B}(r,t)$ (dots denoting t-derivatives), a statement of Birkhoff's theorem. The Weyl method can be redeemed in this case using an auxiliary field as detailed in Reference [7] (with the same fix applied to Lovelock gravity establishing Birkhoff's theorem there in Reference [8]) but using just the spherical symmetry by itself is not enough to establish Birkhoff's theorem. A similar auxiliary field is used in (6), where $b(r) = 1$ is an uninteresting solution to a trivial field equation, yet the function $b(r)$ is necessary to constrain $a(r)$

to its correct value by preventing the collapse of the Lagrangian to a total derivative as in (12) which lacked the $b(r)$ starting field.

One way of viewing the problem with proving Birkhoff's theorem is the focus on the two-dimensional $r - t$ subspace of spherically symmetric spacetimes that are, at least potentially, time dependent. The diagonal metric ansatz does not probe enough of that space to capture the time-independent constraint. A similar problem occurs if we attempt to carry out the procedure on a static, axially symmetric spacetime like the Weyl class of metrics. These typically start with line element

$$ds^2 = -e^{2a(s,z)}dt^2 + e^{-2a(s,z)+2b(s,z)}\left(ds^2 + dz^2\right) + s^2 e^{-2a(s,z)} d\phi^2 \tag{14}$$

for unknown functions $a(s,z)$ and $b(s,z)$ exhibiting cylindrical symmetry (no ϕ dependence). The Weyl method again fails to return a complete set of field equations, in this case because we have started off with the $s - z$ subspace in its (guaranteed) conformally flat form. Here, again, a Lagrange multiplier procedure can be used to restore the 3 independent field equations from Einstein's equation in vacuum but this must be done explicitly.

4. Reissner-Nordström and Magnetic Monopoles

Weyl's method works for extended sources as well as the simpler vacuum solutions provided the sources can themselves be fit into a field-theoretic action in combination with the Einstein-Hilbert action. The gravitational field variables show up in the auxiliary action in the usual way, both through the density $\sqrt{-g}$ and any explicit metric dependence, for example, $g^{\mu\nu}$ in the Lagrangian for a scalar field ϕ: $\phi_{,\mu} g^{\mu\nu} \phi_{,\nu}$ (the method is not available for non-Hilbert stress tensors, making it difficult to use in a cosmology context with fluid stress tensor sources). We can obtain the spherically symmetric spacetime for a charged massive spherical central body by starting with the combined Einstein-Hilbert and E&M action:

$$S = \int \sqrt{-g}\left(R + \sigma F^{\mu\nu} F_{\mu\nu}\right) d^4x \tag{15}$$

where σ is just a constant to set the coupling between gravity and E&M.

Now let's use the Weyl method to find the static, spherically symmetric solutions away from the massive source as in Reference [3]. Start with the ansatz from (6) for the gravitational piece, then the electromagnetic portion reads

$$F_{\mu\nu} F^{\mu\nu} \equiv F^{\mu\nu} F^{\alpha\beta} g_{\mu\alpha} g_{\nu\beta} = 2\left(B^2 - E^2 b(r)^2\right) \tag{16}$$

which depends on the metric used to contract the field strength tensor indices. The starting action is

$$S = \int \sqrt{-g}\left(R + 2\sigma\left(B^2 - E^2 b(r)^2\right)\right) d^4x \tag{17}$$

For a spherically symmetric electric charge source, $\mathbf{B} = 0$ and $\mathbf{E} = E(r)\hat{\mathbf{r}}$. The electric field comes from the $A_0(r)$ term in the vector potential A_μ where the lower form is the relevant one (since the field strength tensor is naturally covariant, $F_{\mu\nu} \equiv \partial_\mu A_\nu - \partial_\nu A_\mu$). In terms of this single non-zero term in the four-potential, $E(r) = A_0'(r)/b(r)^2$ and the Lagrangian is

$$\mathcal{L} = -\sin\theta\left[b(r)\left(-2r + \left(r^2 a(r)\right)'\right)' + \left(2r^2 a(r) b'(r)\right)' + r^2 a'(r) b'(r) + \frac{2\sigma r^2 A_0'(r)^2}{b(r)}\right] \tag{18}$$

Using the Euler-Lagrange equations that come from varying the associated action with respect to $a(r)$, $b(r)$ and $A_0(r)$ independently, we get

$$\begin{aligned} 0 &= 2r\sin\theta b'(r) \\ 0 &= -2\sin\theta\left(-1+(ra(r))' - \frac{r^2\sigma A_0'(r)^2}{b(r)^2}\right) \\ 0 &= \frac{4r\sigma\sin\theta}{b(r)^2}\left(-rb'(r)A_0'(r)+b(r)\left(2A_0'(r)+rA_0''(r)\right)\right) \end{aligned} \tag{19}$$

The first equation is trivially solved by setting $b(r) = b_0$ a constant (that can be set to one by coordinate rescaling). The third equation, simplified using the first, is $2A_0'(r) + rA_0''(r) = (r^2A_0'(r))'/r = 0$. Its solution is $A_0(r) = V_0 - \beta/r$ for constant V_0, the value of the potential at spatial infinity and a constant β that is proportional to the electric charge. With these two in place and taking $V_0 \to 0$, the middle equation reads

$$-(ra(r))' + \left(1+\sigma\frac{\beta^2}{r^2}\right) = 0 \longrightarrow a(r) = 1 - \frac{\alpha}{r} - \sigma\frac{\beta^2}{r^2}, \tag{20}$$

where α is related to the mass of the central body as in the Schwarzschild case. The line element and potential are

$$\begin{aligned} ds^2 &= -\left(1-\frac{\alpha}{r}-\sigma\frac{\beta^2}{r^2}\right)dt^2 + \frac{1}{\left(1-\frac{\alpha}{r}-\sigma\frac{\beta^2}{r^2}\right)}dr^2 + r^2\left(d\theta^2+\sin^2\theta d\phi^2\right) \\ A_0(r) &= -\frac{\beta}{r}, \end{aligned} \tag{21}$$

which is the correct Reissner-Nordström solution. Note that $A_0(r)$ is related to the electric field magnitude, for $b(r) = 1$, by $E(r) = A_0'(r) = \beta/r^2$, the usual Coulomb field associated with a spherically symmetric charge (the covariant zero-component of the four-potential, A_0, plays the role of $-V(r)$ for the usual electrostatic potential $V(r)$).

Let's now consider the spacetime associated with a massive spherical central body with magnetic monopole charge (but no electric charge). All that changes is that we take $\mathbf{E} = 0$ and $\mathbf{B} = B(r)\hat{\mathbf{r}}$, this time with $B(r) = W'(r)/b(r)^2$ for a magnetic monopole potential $W(r)$ replacing $A_0(r)$ from above. Looking at (17), it is clear that the sign associated with the magnetic field is opposite that of the electric case and that ends up introducing a minus sign in the line element that solves the field equations. The Lagrangian is now

$$\mathcal{L} = -\sin\theta\left[b(r)\left(-2r+\left(r^2a(r)\right)'\right)' + \left(2r^2a(r)b'(r)\right)' + r^2a'(r)b'(r) - \frac{2\sigma r^2W'(r)^2}{b(r)^3}\right] \tag{22}$$

In addition to the sign change, there is a factor of $1/b(r)^3$ attached to the potential, as opposed to the $1/b(r)$ in (18). This ends up introducing an essentially irrelevant factor of 3 in the metric's dependence on magnetic charge.

For constant $\bar{\beta}$ associated with the magnetic charge, the line element and potential that comes from the Weyl method applied to (22) is:

$$\begin{aligned} ds^2 &= -\left(1-\frac{\alpha}{r}+3\sigma\frac{\bar{\beta}^2}{r^2}\right)dt^2 + \frac{1}{\left(1-\frac{\alpha}{r}+3\sigma\frac{\bar{\beta}^2}{r^2}\right)}dr^2 + r^2\left(d\theta^2+\sin^2\theta d\phi^2\right) \\ W(r) &= -\frac{\bar{\beta}}{r} \longrightarrow B(r) = \frac{\bar{\beta}}{r^2} \end{aligned} \tag{23}$$

From this, we would conclude that the Reissner-Nordström solution for a magnetic monopole has a fundamentally different structure than the electric monopole case, with $\beta^2 \to -3\bar{\beta}^2$ taking us from one metric to the other. Again it is the minus sign that is important here, that's what changes the structure of the spacetime (in particular, the horizon structure is different between the two).

Einstein's field equations tell a different story—the correct one, of course [9,10]. For the electromagnetic sourcing, we consider the full field equations,

$$R_{\mu\nu} - \frac{1}{2}g_{\mu\nu}R = 8\pi T_{\mu\nu}, \tag{24}$$

and note that the elements of the electromagnetic stress tensor,

$$\begin{aligned} T^{00} &= \frac{1}{2}\left(E^2 + B^2\right) \\ T^{ij} &= \left(\frac{1}{2}\delta^{ij}E^2 - E^iE^j\right) + \left(\frac{1}{2}\delta^{ij}B^2 - B^iB^j\right) \end{aligned} \tag{25}$$

are symmetric in $\mathbf{E} \leftrightarrow \mathbf{B}$, while the Poynting vector contribution, $T^{0i} \sim (\mathbf{E} \times \mathbf{B})^i$ vanishes when considering either field in isolation (this is true even in the extended setting in which magnetic monopoles are incorporated in Maxwell's equation from the start). Then the role of an electric or magnetic monopole in gravity is the same, we have $\beta^2 \to \bar{\beta}^2$ in (21). The field equations give a different result than the Weyl method, so this is another example where the Palais principle is violated.

The problem in this case comes from the move from the electromagnetic action, $\mathcal{L}_{\text{EM}} \sim \sqrt{-g}F_{\mu\nu}F^{\mu\nu}$, to the electromagnetic stress tensor (obtained by Hilbert's procedure),

$$T_{\mu\nu} \equiv \frac{2}{\sqrt{-g}}\frac{\partial \mathcal{L}_{\text{EM}}}{\partial g^{\mu\nu}}. \tag{26}$$

The variational procedure that generates the stress tensor source for gravity from (15) requires that we probe the full metric dependence of the electromagnetic action. In the Weyl method, we probe only a subset and evidently, that subset is too small to reproduce the correct stress tensor structure. Indeed, the electromagnetic piece of the Lagrangian,

$$\mathcal{L}_{\text{EM}} = \sqrt{-g}F_{\mu\nu}F^{\mu\nu} = 2r^2 b(r)\sin\theta\left(B^2 - b(r)^2E^2\right) \tag{27}$$

depends on only one of the metric's two independent functions, a clear warning sign that we will be looking at only a portion of the stress tensor defined by (26).

The situation is similar to the failures described in Section 3 but in those cases, only the gravitational piece was relevant. We know that the line element ansatz (6) used here is enough to capture the spherically symmetric vacuum solution but it is not enough to provide the correct source term. It is conceivable that we could introduce Lagrange multipliers to restore the procedure, as with Birkhoff's theorem or the axially symmetric Weyl metrics. But the ease with which we obtain the correct solution from the full field equations makes the task of finding such fixes unnecessary.

5. Conclusions

Symmetry is a powerful simplifying tool in many settings. In general relativity, the Weyl method makes good use of symmetry observations by reducing the number of degrees of freedom in the Einstein-Hilbert action. The method does not always work and it is important to test solutions obtained in this way by running them through the Einstein field equations. While the Weyl method can be used to correctly obtain the Reissner-Nordström spacetime outside of an electrically charged spherical central body, it fails to produce the correct spacetime when the central body is magnetically charged.

For most applications, the problem with the Weyl method is similar in spirit to the extreme example from Section 3, where the degrees of freedom in the action have been over-reduced, leaving us with no information. That's what happens for the spherically symmetric starting point (11) and the same deficiency occurs when trying to prove Birkhoff's theorem and establish the Weyl class of metrics starting from (14). In each of these cases, the hallmark is a lack of information, the field variables are unconstrained in some way that they should be, according to Einstein's equation. We are left with no information and that lack of information tells us that the error has occurred and suggests a fix.

The monopole case discussed here is different. We are not simply missing information that we suspect should be there. Rather, the information we have is incorrect. The symptom is different but the deficiency is the same, a lack of ability to probe the starting action's full field degrees of freedom. This time, it is the "source" term, the electromagnetic action, rather than the gravitational one that is the culprit.

Conflicts of Interest: The author declares no conflict of interest.

References

1. Weyl, H. *Space-Time-Matter*; Dover: New York, NY, USA, 1951.
2. Boulware, D.G.; Deser, S. String Generated Gravity Models. *Phys. Rev. Lett.* **1985**, *55*, 2656–2660. [CrossRef] [PubMed]
3. Deser, S.; Tekin, B. Shortcuts to high symmetry solutions in gravitational theories. *Class. Quant. Grav.* **2003**, *20*, 4877–4884. [CrossRef]
4. Deser, S.; Franklin, J. De/re-constructing the Kerr metric. *Gen. Rel. Grav.* **2010**, *42*, 2657–2662. [CrossRef]
5. Palais, R.S. The Principle of Symmetric Criticality. *Commun. Math. Phys.* **1979**, *69*, 19–30. [CrossRef]
6. Deser, S.; Franklin, J.; Tekin, B. Shortcuts to spherically symmetric solutions: A cautionary note. *Class. Quant. Grav.* **2004**, *21*, 5295–5296. [CrossRef]
7. Deser, S.; Franklin, J. Schwarzschild and Birkhoff a la Weyl. *Am. J. Phys.* **2005**, *73*, 261–264. [CrossRef]
8. Deser, S.; Franklin, J. Birkhoff for Lovelock Redux. *Class. Quant. Grav.* **2005**, *22*, L103. [CrossRef]
9. Bais, F.A.; Russell, R.J. Magnetic-monopole solution of non-Abelian gauge theory in curved spacetime. *Phys. Rev. D* **1975**, *11*, 2692–2695. [CrossRef]
10. Lee, K.; Nair, V.P.; Weinberg, E.J. Black holes in magnetic monopoles. *Phys. Rev. D* **1992**, *45*, 2751–2761. [CrossRef]

MDPI
St. Alban-Anlage 66
4052 Basel
Switzerland
Tel. +41 61 683 77 34
Fax +41 61 302 89 18
www.mdpi.com

Symmetry Editorial Office
E-mail: symmetry@mdpi.com
www.mdpi.com/journal/symmetry

www.ingramcontent.com/pod-product-compliance
Lightning Source LLC
Chambersburg PA
CBHW051916210326
41597CB00033B/6160